PHYSICS OF THE FUTURE

A Classical Unification of Physics

by
Thomas G. Barnes

Institute for Creation Research
El Cajon, California

PHYSICS OF THE FUTURE
A Classical Unification of Physics

Copyright © 1983 Thomas G. Barnes

Published by
Institute for Creation Research
2100 Greenfield Dr.
El Cajon, California 92021

Library of Congress No. 83-081788
ISBN 0-932766-09-9

Cataloging in Publication Data
Barnes, Thomas G.
   Physics of the future: a classical unification of physics.
   1. Physics.   I. Title
      530

Printed in the United States of America
ISBN 0-932766-09-9

Foreword

In the present climate of opinion, there is nothing harder than to go back; to perform a regress (as C.S. Lewis called it in *The Pilgrim's Regress*); to admit that, on such-and-such points, our predecessors were right after all. But such a regress may be just what is needed; as Lewis remarked in another place, if the clock is wrong it may need to be turned back.

This book is, in one sense, such a regress; it goes back to pick up a trail which was lost early in this century.

In the nineteenth century, it will be recalled, Maxwell and others had laid a solid foundation of electrodynamics. Among the points which emerged was the fact that electromagnetic effects are propagated at a speed equal to that of light. The common formulations of electromagnetism are based on experiments in which the sources were stationary or moving at speeds much less than that of light. If the sources should be moving at speeds approaching that of light, it would not be surprising if some effects were to seem different. In the same way, when speeds exceeded that of sound, it was necessary to look afresh at some problems in aerodynamics; but nobody suggested that we should throw out the whole of the subject and start over.

In a parallel development, from about the middle of the nineteenth century onward, the constitution of matter was studied. The result was a general agreement that matter is granular in nature; that it is to be considered made up of particles; that these particles are many or all of them electrically charged, and that the electrical interactions determine the properties of matter.

Then, around 1920, a change took place. Anyone can see it by looking through *Physics Abstracts:* how about that time there was a great decrease in studies of the physical actions of real objects, macroscopic or microscopic, and an increase in mathematical manipulation and in shuffling equations with little apparent physical content. It may be that the theory of relativity, which apparently did not attract wide notice until around 1920, was to blame, along perhaps with other developments; maybe the climate of opinion following the Great War had something to do with the matter. Anyway, in the author's opinion, and that of others who could be cited, the change was not a good one.

This work, then, represents a return to the method which takes classical electrodynamics, develops it further if necessary, and applies it to microscopic problems. Of course, the work is more than a regress. Experimental results are now known which were unavailable to the founders of electromagnetism. Such results may help to solve problems which were formerly difficult; on the other hand, they may cause new problems.

First of all, then, the situation in electrodynamics in which sources are moving at high speeds, i.e. approaching that of light, is investigated. The author, drawing on his experience in electronics, calls the peculiar effects which happen feedback. Thus, a moving electric charge causes a magnetic field which is changing because of the motion. The magnetic field, in turn, is changing, and thus causes an electric field which acts back on the source.

This idea is not claimed to be new; Abraham, Lorentz, and others knew of it. But the author has applied it in a more systematic way than seems to have been tried previously.

Now the feedback force on a moving charge is such as to

oppose any change in its motion. From the viewpoint of electromagnetism, this is just Lenz's law. But from the viewpoint of mechanics, the moving charge has inertia; it has mass. The view is taken that, since all matter is composed of electric charges, all matter will have mass for that reason; and that, in fact, inertia is an electromagnetic effect.

Since the arrangement of the electric and magnetic fields becomes different as speeds approach that of light, the inertia becomes different under such conditions; and this difference may be called an increase in mass. The results for that effect are identical with, or similar to, those developed in relativity. But no appeal is made to observers wandering around with flashlights and rubber yardsticks; the effects are physical ones and can be discussed by physical arguments. Our business, as Newton said, is with the sensible causes of the phenomena.

Incidentally, it is worth noticing that the experiments which show an increase of mass with speed have all been on charged elementary particles. There are no such experiments with neutral macroscopic objects.

The notion of mass arises also in connection with gravitation; and it is natural to ask whether the two aspects are connected. The answer proposed here is: "Yes." It is proposed that, as for electrical neutral objects, i.e., containing equal amounts of positive and negative charge, the forces of attraction and repulsion cancel almost exactly. But, due to a slight non-linearity, the force of attraction always preponderates very slightly; and this preponderance of attraction is gravitation.

When fields are mentioned, and still more waves, the question arises: "in, on, or of what?". In effect, the present treatment allows for an ether, without becoming committed to any of the models of the ether which proliferated in the last century. Certainly some medium - or call it what you will - is needed; to talk of waves or fields in nothing at all is nonsense; it is like talking about the color of a non-existent house.

It is generally considered impossible to get objects to move in translation at speeds greater than that of light; and this work does not dispute that idea. But it is suggested that the peripheral speed of a rotating elementary charge, in particular, may not be under such a restriction. So the result could be magnetic forces between particles which would provide an attraction stronger than the electrostatic repulsion. Such forces may be the short-range forces which hold atoms, molecules, and macroscopic objects together. If this view is correct, here is unification comparable with that which was achieved when the connections between electricity and magnetism were discovered. Some tentative models are worked out on this basis.

Our views on the structure of atoms have been much influenced by the study of radiation; so it is natural that that subject be considered. Commonly the quantum theory, and the notion of photons, are invoked. Here the notion that light in propagation is to be considered a stream of photons is rejected. The wave theory, which is based on abundant evidence from such phenomena as interference, is adequate to deal with propagation. Nor are the observations claimed to compel the idea of photons really all that compelling. After all, the storm may blow down one tree in the forest; but nobody says that an *anemon* collided with that tree only. When we study the emission and absorption of radiation, we are studying the action of materials in emitting and absorbing radiation, not the nature of light itself.

Of course, it is impossible fully to develop such a viewpoint (which has also been suggested by Ives, Poincaré, and others) within the compass of this work. But it is possible to suggest how to begin.

In commending this work, then, to the reader's attention, may I make two concluding remarks? In the first place, the author realizes that he has left questions unanswered, i's undotted and t's uncrossed. To some extent, what is offered here is a programme for further studies. Such a state of affairs is natural; only rarely is it possible for one man to begin a line of investigation and bring it to a complete conclusion. Indeed, it may be questioned whether such a thing would even be desirable. It may be that a work in which many have taken a hand will usually prove better.

The second point is this. The author has proposed such things as models of neutrons and atoms. These are treated as models in the sense in which a model ship is a model. Now even if it should happen later that these have to be modified, they will still be at least analogues, in the sense in which an electrical network is an analogue of some mechanical progress. Now the mathematical representations, void of physical content, which are so much in vogue, are surely just analogues, in this sense. So there is nothing to be lost by considering the ideas proposed here, and seeing how far they may be carried; and there may be much to be gained.

Harold L. Armstrong
Kingston, Ontario, Canada

Preface

This book exposes serious problems with Einstein's theories of relativity and with certain areas of quantum theory. It rejects the basic postulates of relativity and quantum theory and attempts to develop a foundation for modern physics employing an expansion of the concepts of classical physics.

It is not just a matter of substituting one type of treatment for another. It now appears that the classical approach, when it is completely developed, may simplify modern physics. For example, it may be possible to unify *all* forces in physics into only two fundamental forces, namely electric and magnetic forces. That contrasts with the present physics classification of five fundamental forces: gravitational force, weak force and strong force (associated with the atom and nucleus), and electric and magnetic forces. An extension of the classical approach shows promise of reducing each of these extra three forces (the gravitational force, the weak force, and the strong force) to electric and magnetic forces.

In the extension of classical physics, as an alternative to relativity, a new feedback process is postulated. This provides new physical phenomena that restore the concepts of absolute time and Euclidean geometry. New interpretations of experiments are advanced to support that position. From there one is able to show the possibility of speeds exceeding the speed of light in certain types of rotational motion.

If these new concepts are anything like as promising as they now appear to be, physics of the future is going to be mighty exciting for young physicists who are open to challenge.

El Cajon, California   Thomas G. Barnes
June 1983

Acknowledgements

Those familiar with the present status of modern physics and cosmology will surely realize the monumental task undertaken in this book. The departure from the views of the *elite* is almost total and the possibilities for error are legion. It tempts one to include the statement: The views expressed herein are those of the author and not necessarily those of the publisher nor the acknowledgees.

Be that as it may the author has received encouragement from many sources, far too many to enumerate. This has been going on through the years with his acquaintances, particularly since he began to publish papers that form part of the background for this book. One of the most rewarding experiences has been the encouragement from his students through the years. They always seemed to delight in having a part in the intellectual game of trying to simplify and unify the very foundations of physics. It is not possible to acknowledge those students individually but this is meant to express my thanks to them all for their encouragement.

Although many scientists have read some versions of the manuscript and have my gratitude, the personal acknowledgements will have to be limited to the following who made specific contributions that helped make this book possible: Dr. Harold S. Slusher has been my closest colleague and a continual consultant over a broad spectrum of science. Harold Armstrong has for years given many valuable suggestions and references and has graciously agreed to write the Foreword. Dr. Richard R. Pemper made a major contribution through his original work that can be found in his Master of Science Thesis at the University of Texas at El Paso. I can never say enough about the inspiration he has given me since I had him as a graduate student. Raymond J. Upham, who obtained his M.S. Degree from UTEP, made a substantial contribution through his collaborative work on a paper that provides an alternative to Einstein's General Theory of Relativity. Francisco Salvador Ramirez Avila IV made a real contribution through his Master of Science Thesis at UTEP which showed an excellent classical means of deducing the presumed observational evidences for Einstein's General Theory of Relativity. The author is particularly thankful to Dr. Russell Humphreys for bringing to his attention the electric reaction force of an accelerated current and deriving the illustrative example which is included in this book as Appendix II.

The author is grateful to Dr. Henry M. Morris, President of the Institute for Creation Research and Editor of the Publications Division, for making the decision to publish this book. Others on the staff of the Institute who have made direct contributions to the book are: Donald H. Rohrer, Business Manager, who did the expediting of the publication; Mrs. Becky Nichols who patiently and efficiently took the manuscript and various revisions and corrections and put it on the word processor and arranged it into camera ready form and followed it through the final stages of publication; and Marvin Ross, who did the art work. I am grateful to all of the above for their fine contributions to the book.

The one person who has really pulled me through, however, is my wife Elizabeth (Libby). She took burdens off of me and encouraged me to stay at the writing. She typed and assembled an innumerable number of versions of the manuscript. My gratitude goes to her and to our children and their families, all of whom have given the author wholesome encouragement in this and his other scientific efforts.

Contents

CHAPTER 1
Introduction

1-1 Philosophical indoctrination in physics

Technology has advanced at an ever increasing rate. Within a life span there has been the transition from hand cranked calculators to telecomputing systems that have revolutionized business and industry. There has been the transition from the horse and buggy to spacecraft. Surprisingly some recent technological advances turn out to be ingenious applications of old fundamental principles in physics. The trajectories of spacecraft are computed electronically but the physics goes back to fundamental physics known in the eighteenth century. The physics is old. The technology is new.

While the advances in technology have been continuously upward, it is a matter of opinion as to which way the curve slants on the rate of real advances in the fundamentals of modern physics. Dr. C.S. Cook made a study of the Nobel Prize winners in physics since about 1940. He concluded that most of them had been to older physicists whose prize winning work was many years prior to the time of their award. The conclusion was that progress in making real fundamental discoveries in physics is on the downgrade.

The philosophical pendulum in physics has swung too far to the left. The spirit of individual creative potential has all too frequently been discouraged. The young physicists are overwhelmed by an excessive emphasis on complex mathematics and a minimum of physical reasoning. The whole structure of modern physics appears to thrive on the very opposite of common sense.

It is a hazardous thing in science to attempt to define *common sense*, but the normal person has some concept of what the term means. It certainly would imply clear and rational thinking, and in physics would associate physical effects with physical causes. The layman might expect the prevailing views in the theoretical foundations of *modern* physics to be based on a measure of common sense. Not so! The elite position taken by most theoreticians in modern physics is to leave no place for common sense in the development of the basic principles.

In the early days of physics when physics was known as *natural philosophy*, common sense was at a premium. That type of physics is now called *classical physics*. The physics of Isaac Newton, Michael Faraday, James Clerk Maxwell, Lord Rayleigh, Lord Kelvin, Karl Gauss, and Ernest Rutherford is classical physics. The space flights achieved by NASA are illustrations of applied classical physics.

The underlying theoretical principles of all *modern* physics are: 1) Einsteinian relativity and 2) quantum mechanics. Those principles are not common-sense principles. While acknowledging many successes with those principles in modern physics, the author contends that there must be a better way, one that associates physical effects with physical causes, a common-sense approach.

1-2   The need for a change
There has now been several generations of indoctrination in a philosophical acceptance of "no need to be concerned with cause and effect relationships." The result is an unquestioning acceptance of the philosophy of: *putting the cart before the horse*, assuming that one can start with mathematics and explain the physics rather than starting with physical reasoning and employing mathematics to help illucidate that reasoning. This inverted type of logic frequently leads to self-

contradictions and sheer nonsense.

Herbert Dingle exposes that type of nonsense in the writings of the noted astronomer Sir Fred Hoyle. Hoyle goes beyond the bounds of common sense in the use of mathematics. Dingle quotes Hoyle as follows:

> *"To the scientist war starts because human behavior is representable in terms of mathematical equations possessing discontinuous solutions."*

Dingle then explains:

> *"This must not be dismissed as a humorous wisecrack: Hoyle, and others of his type, really believe that this is so. They were not necessarily born with a deficiency of common-sense: They have exceptional mathematical ability which has been mistaken for exceptional intelligence, and have been so trained that their normal intelligence, has expired through desuetude: much mathematics has made them - what they are."* [1]

It is the courage and great scholarship of scientists like Herbert Dingle that encouraged the author to question the Einstein theories of relativity, and to develop alternatives. Many other scientists have now joined Herbert Dingle's bandwagon and openly question Einstein's theories of relativity. Some question other phases of modern physics. There is a need not only to give alternatives to the relativity theories but also to give alternatives to the other foundational principles of modern physics. That is far too much for one individual to achieve. The aim of this book is to present illustrative examples of plausible alternatives to be considered. It will be seen from these illustrations that when one foundational principle is changed it also requires changes in the other principles. Hence there is a need to eventually revamp the entire framework of modern physics.

1-3 Revolutionary new alternatives

The alternatives presented in this book should be considered as tentative and incomplete. Rather than holding off until they are completely developed, the author has chosen to include them in their unfinished state. The purpose is to illustrate new ways of physical analysis in these areas of physics. It is hoped that they will serve as encouragement to the physicists who wish to be more innovative and original. Their work may determine the soundness of physics of the future.

The author owes a great deal to his fine graduate students who have been engaged in the search for alternatives to Einstein's relativity. Much of the background of this book was developed through joint papers with the author's students and colleagues.

This book includes observational and experimental evidences that appear to invalidate Einstein's theories of relativity. It provides some alternatives to his relativity and other concepts and principles in modern physics. One of the alternatives to relativity is a feedback postulate. This postulate also made possible the development of new models for the electron, proton, and neutron. In those models the proton is smaller than the electron and the neutron consists of a spinning electron and proton. This approach provides a logical and consistent means of eliminating some of the supposed problems with classical physics.

It is shown that an elementary charge can have a rotational motion in which the peripheral speed exceeds the speed of light. This is an extremely important development. It provides a classical explanation of the *strong force* in the atom, a magnetic force that exceeds the Coulomb electric force.

An electric explanation is given for the dynamic reaction force in Newton's third law. Inertial mass is shown, for speeds much less than the speed of light, to be an electric property. The inertial mass for higher speeds appears to be a logical extension of this electric theory, but that problem remains to be solved as time permits.

An electric theory of gravitation is introduced. Gravitational mass is developed as an electric property. The theory includes the postulating of an overload factor at or near the surface of the electron and of the proton.

Alternatives are proposed to some areas of quantum mechanics. The atomic emission and absorption of light are assumed to be resonance phenomena within the atom. The *particle* property of light is rejected, assuming that light is only a wave phenomenon of a classical type.

1-4 Advantages of the new approach

It is the author's contention that drastic alternatives, such as these, are needed throughout the whole foundations of modern physics. There are still many important unsolved problems in modern physics. The long sought unified field theory has eluded all efforts of the best physicists in the past. Hopefully these alternatives will provide new incentives and a fresh start from which a new generation of young scientists can develop a more excellent formulation of physics that incorporates common sense and associates the effect with the cause.

The question may be raised as to how one can ignore the experimental evidence upon which modern physics is presumably based. One answer is that most experiments are not so definitive as one might think. The interpretation of an experiment is often dependent upon many assumptions related to the experiment itself. In some instances an alternate interpretation of the experimental results makes more sense. For example, *instead of interpreting the extended lifetime of very high speed muons as evidence that time itself runs slower (a relativity notion) one can interpret the results as evidence that the "stability" of this particle is increased when it has more energy. In other words, instead of time itself running slower, the accuracy of the clock is altered.* This alternate interpretation might also provide clues to the unsolved problem of the cause of radioactive decay.

In some instances the scientist has been more interested in "promoting" his theory than in strict adherence to experimental evidence. A recent article that reviewed some noted experimental work pointed out that Einstein claimed that experimental evidence of one particular experiment *agreed* with his theory, whereas the best analysis of the experimental data showed that it fit another theory better than his theory.[2]

What is involved here is more than science. It is also philosophy. The philosophical base for modern physics is radically different from the philosophical base for classical physics. In modern physics Einstein's relativity has nurtured the philosophy of *relativism*. Modern physics is also founded upon *statistical probabilities and chance, as opposed to cause and effect*. It incorporates the uncertainty principle into quantum theory. Even though the statistical approach has been quite successful in certain areas of physics, the elevation of non-causal concepts to a philosophy of science has been perhaps the greatest source of degradation in science. It has led to the philosophy that "anything can happen if given enough time." Eddington stated that a group of monkeys at typewriters could by time and chance produce a perfect copy of the Encyclopaedia Britannica.

When that kind of philosophy has displaced real science in our "centers of learning" the opportunities are unlimited for an independent thinker who has some common sense. It is the author's hope that this book will encourage young independent scientists to take heart, abandon that nonsense, be original and fruitful in science, and enjoy a more wholesome philosophy.

References

1. Dingle, Herbert 1972. Science at the crossroads. Martin Brian & O'Keeffe, London, p. 129.
2. Cushing, James T., Electromagnetic mass, and the Kaufmann experiments, *American Journal of Physics* 49 (12):1147.

CHAPTER 2
Rejection of Special Theory of Relativity

2-1 Challenge from an expert

Einstein made famous the German word *gedankenexperiment*, which has the English translation *"thought experiment."* His thought experiments were strictly hypothetical constructs, fancied experiments of an impractical nature. Nevertheless, he considered the outcome to be something that he could predict. He used thought experiments to illustrate his relativity and to dramatize hypothetical relativity applications such as the prolonging of a person's lifetime by fast travel.

Dr. L. Essen, a famous English scientist who strongly refutes Einstein's theory of relativity, points out that "thought experiment" is a contradiction in terms. Thought is not an experiment. Dr. L. Essen is one of the world's leading authorities on time measurements. This author's interest in developing an alternative to Einstein's special theory of relativity was enhanced by correspondence with Dr. Essen. The following excerpt from one of his letters shows the scientific need for challenging Einstein's relativity:

20-9-77

Dear Professor Barnes,

. . .As a practical physicist concerned with time measurement I knew that Einstein's thought experiments were car-

ried out incorrectly and gave nonsensical results; and my
interest in standards and measurements enables me to
point out with confidence some of the other mistakes in
Einstein's paper.

...However, my aim in trying to demolish Einstein's theory
was to make room for new theories and in particular for a
`rational` theory of space and propagation. As you are no
doubt finding it is extremely difficult to get any unconventional theory published or taken notice of. However, we
must keep trying.

                                        Sincerely yours,

                                        L. Essen

2-2 Problems with relative time

Einstein had two theories of relativity. The one of interest
here is called the special theory of relativity. According to
special theory of relativity time runs slower in a moving frame
of reference, such as a high speed rocket. This slowing down
of time is called *time dilation*. Einstein's thought experiment
for time dilation was to have one twin ride off in a high speed
rocket and the other remain at home. When the high speed
traveler returns home he is younger than the stay-at-home
twin. The late Herbert Dingle, a former president of the
royal Astronomical Society and the author of two books on
relativity, pointed out that the presumed time dilation leads
to a logical contradiction, requiring a clock to run both slow
and fast at the same time. He based his argument on the additional tenet of special relativity that there is no absolute fixed
frame of reference in space. Any inertial (non-rotating or non-
accelerating) frame may, according to the rules of special
relativity, be arbitrarily chosen to be a "fixed" frame, for
convenience in solving the problem. But Dingle shows that if,
as Einstein contends, either of two frames may be taken as
fixed or moving with respect to the other, this logically demands the impossible, namely that a clock can run both fast
and slow at the same time.[1]

It appears that the classic illustrations of time dilation, such as the twin paradox and another famous one known as the muon experiment, can not really stand up to the probing of Dingle's analysis. Furthermore, an atomic clock experiment that is frequently cited as proof that moving clocks run slower has recently been reinterpreted otherwise. The experiment involves flight around the world with atomic clocks and the conclusion that the fastest moving clocks did run slower. Dr. L. Essen, who is the inventor of the atomic (Caesium) clock, has taken *all* of the data from that experiment and shown that it does not prove that there was any time dilation.[2] There is no one more qualified to challenge that experiment than Dr. Essen.

2-3 Problems with space contraction

According to special relativity, space contracts in a moving frame of reference, such as a high speed rocket. The space contraction is in the direction of motion, with no contraction in the direction at right angles to the motion. Note that space contraction and time dilation are said to occur in the "moving" frame of reference, but no time dilation nor length contraction in the "fixed" frame of reference. Time and space are said to be relative, dependent on relative motion with respect to an "observer." These relativity effects are supposed to be so small that they are essentially undetectable until the speed is very great, approaching the speed of light, three hundred million meters per second. No actual rocket ever approaches such speeds. So direct checks of time dilation and length contraction are not easy to achieve. As previously mentioned, the presumed experimental evidence for time dilation is questionable.

The Einstein *space* contraction should not be considered to be synonymous with the contraction of a rod, such as a metal meter stick. Space length and rod length are not the same. For example, a metal rod's length may expand when its temperature is raised, but space length is independent of temperature. Lorentz developed a relativistic equation for length contraction of a *rod* as a function of speed, before Einstein's 1905 paper on special relativity. Lorentz considered this rod contraction to be due to a physical cause, namely a force exerted on the rod by the *ether* through which it was moving. Maxwell had proposed ether as a medium that filled all space and provided the mechanism for the propagation of light through space. Lorentz retained that ether concept.

Note that Lorentz had a physical reason for the contraction of the rod, whereas Einstein never gave a physical reason for his proposed space contraction. He set the trend in physics and cosmology of excluding the logic of cause and effect. The emphasis shifted from physical reasoning to mathematical manipulation.

Einstein "borrowed" Lorentz's transformation equations for time and length but put a different interpretation on them. In his original papers Einstein did not acknowledge the prior work of Lorentz in deducing those equations. Today scientists acknowledge the prior work of Lorentz and call those equations the Lorentz-Einstein transformations, or simply the Lorentz transformations. One might extend the physical reasoning of Lorentz, and say that those equations imply the alteration of the rate of a clock (not time itself) by the physical effects on the clock due to motion. That is different, however, from considering time as relative and space as relative, as Einstein perceived them. Einstein did not consider rod length shortening with speed to be due to a physical force on the rod. He considered space and rod contraction to be relative, not the same for all observers. As viewed by an observer in another frame moving with another speed the same rod would, according to Einstein, have a different length. To him, space contraction adapts itself to the motion of the observer. Length and time each are relative quantities, having various values depending on the relative motion with respect to an "observer" as prescribed by certain mathematics, not by physical causes.

There has never been any direct experimental evidence for the space contraction predicted by the special theory of relativity. In his book, *Special Relativity,* Albert Shadowitz states:

> It is an amazing fact that there does not seem to exist any direct or simple experimental verification of the Lorentz-Einstein contraction.[3]

2-4 The ether controversy

Maxwell never established a satisfactory explanation of the "medium" that carried electromagnetic waves. He gave it an imposing name, "luminiferous ether," but some of the properties he ascribed to it seem to fail to agree with experimental results. He thought of the medium as fixed in space, and filling all space, and having mechanical properties. He considered that medium to be an absolute frame of reference with respect to which all motion can be referred.

Maxwell proposed a theoretical way of detecting the presumed absolute reference frame, the ether. He suggested a method of measuring the relative velocity of this medium with respect to the earth as it drifted through absolute space. The instruments of his day were incapable of measuring the small time intervals expected, but with the development of the Michelson interferometer, scientists were able to make indirect measurements that might evaluate time to the desired accuracy. The crux of the experiment was to measure the time required for a light signal to make a round trip a given distance parallel to the ether stream and compare it with the time required for a light signal to travel the same round-trip distance perpendicular to the stream. A simple illustration will show that we would expect a difference in these two times.

Suppose, for simplicity, that instead of a light signal in an ether stream, we illustrate with a boat traveling in a stream of water. Let the speed of the boat with respect to the water be 5 miles per hour. Let the stream of water be moving 3 miles per hour with respect to the bank. Now find the time required for the boat to make a trip 4 miles upstream and the 4 miles back

downstream, as compared with the time to go 4 miles across
stream and the 4 miles back. Referring to Fig. 2-1 we note that
the ground speed of the boat across stream is 4 miles per hour.
Hence, it takes 2 hours for the round trip across stream. The
boat's ground speed upstream is 5-3=2 miles per hour, requiring 2 hours to go 4 miles upstream. Its ground speed
downstream is 8 miles per hour, requiring 0.5 hours to return,
making the total round trip time 2.5 hours. Thus we see that a
round trip parallel to the stream takes more time than a round
trip perpendicular to the stream, although the distance is the
same. In our illustration the difference in time required for the
two round trips is 0.5 hour. It is obviously possible to work
backward and find the speed of the stream when the time difference, the boat speed, and the distance are known.

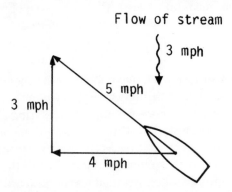

Flow of stream

Fig. 2-1 Time for boat to travel 4 miles upstream and back is greater
than the time to travel 4 miles across stream and back.

In the famous Michelson-Morley experiment, light signals
were employed. An ether stream was assumed to be flowing
past the earth as it moves through space. Light signals traveled
down and back Michelson's interferometer paths that were at
right angles to each other. The astounding result of this experiment was that the "predicted" time difference was not
detected even though the Michelson interferometer was sensitive enough to detect the expected phase shift. The phase shift

was to be interpreted as a measure of time difference in the travel times. Scientist Harold Armstrong pointed out in a letter to the author, that the usual argument for the Michelson-Morley experiment is definitely wrong on one point. "For it mixes a ballistic idea: time of flight, with wave matters: interference."

2-5 Alternate interpretations of the medium

The null effect in the Michelson-Morley experiment seemed to indicate that one can not measure the rate of drift of the earth through the assumed ether which was supposed to be fixed in space. Lorentz maintained that this did not prove there was no ether, but simply that one could not detect the ether by this method. His equations for length contraction and time alteration indicated that these effects would precisely cancel out any measurement of the drift rate through the ether and make it undetectable even though the ether was there.

Einstein is supposed to have taken his clues from the Michelson-Morley experiment in the formulation of special theory of relativity. Burniston Brown has searched the original literature and shown that there is a question as to whether or not Einstein actually did consider the Michelson-Morley experiment when he first developed his theory. Be that as it may, most of the textbooks imply that it was the Michelson-Morley experiment that was foremost in Einstein's mind when he came up with his special theory of relativity. Nevertheless, Einstein did reject an *absolute* frame of reference and the ether theory. Of particular interest in this connection is his second postulate, namely *the speed of light is constant in all inertial frames*. Most of the textbooks give credit to the Michelson-Morley experiment for the demise of the ether theory and list additional experiments as supporting evidence for rejecting the ether medium.

Now we are not sure that those experiments did prove there is no ether. Note for example this quote from the 1975 edition of J.D. Jackson's *Classical Electrodynamics:*

It seems clear that most of the early evidence for the second postulate is invalid because of the interaction of the radiation with the matter through which it passes before detection.[4]

He refers to the extinction theorem. The extinction theorem states that if an electromagnetic wave (light) traveling in a vacuum with speed c enters a dispersive medium, such as the atmosphere, that wave will be cancelled and replaced by another wave propagating with a velocity of that medium. That is to say the incident wave will be extinguished and replaced by another wave. However, Jackson considers a later experiment at CERN, Geneva, Switzerland, to support the second postulate. Another scientist, Wallace Kantor, in his book *Relativistic Propagation of Light*, refutes the validity of the claims made from that CERN experiment.[5]

What is more significant, however, is that there is still a strong possibility that there is a light-bearing medium, a preferred frame of reference of some type. The null effect of the Michelson-Morley experiment does not rule out other possible explanations. In Sec. 12-4 the Sagnac experiment will be cited as evidence that there is a preferred frame of reference with respect to rotation. Insofar as that experiment is concerned one might assume that the preferred frame of reference is the laboratory, containing a local medium of some sort that propagates the light. That medium would also yield the null result in the Michelson-Morley experiment. One should not totally reject the concept of some light bearing medium.

2-6 Scientists who reject Einstein's relativity

It has already been noted that the famous scientists of our day Herbert Dingle, L. Essen, G. Burniston Brown, and Herbert Ives all rejected Einstein's relativity. Other famous scientists who rejected it were H.A. Lorentz and W. Ritz who was developing a rigorous alternative theory until the time of his death at a very early age. Perhaps the most noted dissenter was Ernest Rutherford, the father of nuclear physics. The following quote from G. Burniston Brown's excellent paper *"What's wrong with relativity?"* Institute of Physics and Physical Society, pp. 71-77, March 1967, illustrates how emphatically Rutherford rejected that theory.

> When Wilhelm Wien tried to impress Rutherford with the splendours of relativity, without success, and exclaimed in despair "No Anglo-Saxon can understand relativity!", Rutherford guffawed and replied "No! They've got too much sense!"[6]

Vannevar Bush, President Franklin D. Roosevelt's science adviser, and head of the highly successful Office of Scientific Research and Development during the time of the development of radar, the atomic bomb, etc., is reputed to have considered Einstein's relativity to be an educational barrier to the development of good scientists. The following quote is from a letter from Dr. Elmore E. Butterfield to Dr. John A. Steffens and recorded in the 1979 book *The Einstein Myth and the Ives Papers, A Counter-Revolution in Physics.*

> On more than one occasion Vannevar Bush has emphasized the fact that our system of education is not producing fundamental thinkers of the caliber of J. Willard Gibbs, H. Helmholtz, H.A. Lorentz or H. Poincaré. It is difficult to see how we can produce fundamental thinkers when our teachers cannot detect the fallacies in Einstein's theories, paradoxes and postulates, but instead rush to climb aboard the Einstein bandwagon where further straight thinking becomes impossible.[7]

The following quote from that same book is an excerpt from a letter from Herbert Ives to Dr. Elmore E. Butterfield.

> Apropos of your characterization of Einstein, I think of him as the great paradox swallower, e.g., the velocity of light is independent of the velocity of the source, and also shares the velocity of the source; light is waves and also is

particles. His technique for solving a problem is always to say that both of two contradictory explanations are true.[8]

In *The Listener*, 26 July 1973 under LETTERS, G. Burniston Brown states that:

Practicing physicists and astronomers who know some history of science do not accept "Relativity" and even a distinguished theoretician, Leon Brillouin is calling, in his book *Relativity Re-Examined,* for a "Painful and complete re-appraisal" which "is now absolutely necessary."[9]

2-7   Inconsistent acceptors of Einstein's relativity

Scientists actually run into situations where there seems to be no viable alternative to absolute space, with reference to which all motion takes place. This is a direct contradiction to Einstein's relativity. It is interesting to see how those who ride the Einstein bandwagon handle such situations. In his book, *Theoretical Physics,* George Joos states that there is one situation where there seems to be absolute space. However, he shows his strong support for Einstein's relativity by stating that the "assumption of an absolute space is *superfluous* and in contradiction with experience"[10] insofar as uniform motion is concerned. The contradictory situation relates to accelerated motion and especially to rotational motion. He is not sure how to explain the contradictions:

But the situation is very much more difficult for frames in accelerated motion, e.g. rotation, the phenomena accompanying rotation do seem to point to the existence of an absolute space, for which Newton's second law is valid: the occurrence of inertial forces, such as centrifugal or coriolis forces, indicates that bodies are in accelerated motion with respect to such a space.[11]

He then introduced some conjecture:

But is it not more advisable to say that the accelerated motion takes place with respect to the remaining masses in the universe?

That did not solve the contradiction. Einstein emphatically denied that there is any absolute reference system, whether it be called *ether* or *masses*. There is no "preferred" frame of reference, according to Einstein.

The surprising fact is that modern day cosmologists and many physicists who claim that Einstein's relativity is a fundamental principle of physics conjure up an absolute frame of reference whenever it suits their particular need. How does such a relativist justify the use of an *absolute* frame of reference? It is a matter of rejecting that portion of relativity which does not fit the theory being promoted. Present day cosmologists claim that special theory of relativity does not hold for large scale phenomena. The fact that Einstein never put any such restriction on it is not mentioned.

Cosmologists use microwave radiation that is presumed to "bathe" the universe, as an absolute frame of reference. In his book *Astrophysical Concepts* Martin Harwit states:

> It is interesting that the presence of such a radiation field should allow us to determine an absolute rest frame. . . *special relativity is really only meant to deal with small-scale phenomena. . . phenomena on larger scale allow us to determine a preferred frame of reference. . .* (emphasis added).[12]

Regardless of any supplementary explanations of such a "flip flop" from espousing to rejecting Einstein's relativity, here is another illustration of contradictions that relativists willingly accept. Is it any wonder that Vannevar Bush classified that state of mind as one "where further straight thinking becomes impossible."

2-8   Conclusion

In spite of the popularity of Einstein's theories of relativity, his special theory of relativity and his general theory of relativity, they are not as universally accepted as one would be led

to believe in the classroom and technical and popular literature. That popularity has been kept afloat partly by excluding
from the dominant literature the valid challenges mounted by
many great scientists through the years. The fact that many of
the equations associated with those theories are successful
equations in physics does not mean that Einstein has an ex-
clusive claim on them nor that his interpretation of them is
without contradictions.

The presumed experimental verification of Einstein's special theory of relativity and its time dilation and length contraction are now much less credible than had previously been
acclaimed. In fact, there is not a single so-called verification
of this theory that is without some credible alternative interpretation. His postulate of no preferred reference frame, no
absolute space, does not jibe with the results of experiments
such as the Sagnac experiment and even the common experience of rotational acceleration with reference to a "fixed"
reference.

The logical fallacies in Einstein's interpretations as pointed
out by Herbert Dingle, no matter how well suppressed, appear to be real fallacies. Even his followers in modern cosmology choose to abandon his special theory of relativity
when it does not fit their particular theory of the day. There
are cogent reasons for rejecting Einstein's relativity. The important job ahead is to develop valid and fruitful alternatives.

References

1. Dingle, Herbert, Science at the crossroads. Martin Brian &
   O'Keeffe, London. 1972, p. 17.
2. Essen, L., Personal letter to T.G. Barnes. Sept. 20, 1977.
3. Shadowitz, Albert, Special relativity. W.B. Saunders, 1968,
   p. 168.
4. Jackson, J.D., Classical electrodynamics, second edition.
   John Wiley, 1975, p. 512.
5. Kantor, Wallace, Relativistic propagation of light. Coronado

Press, Lawrence, Kansas, 1976, p. 115.

6. Brown, G. Burniston, What's wrong with relativity. Institute of Physics and Physical Society, March 1967, pp. 71-77.

7. Turner, Dean and Richard Hazelett, The Einstein myth and the Ives papers - a counter-revolution in physics. The Devin-Adair Co., 1979, Part II. The Ives Papers, p. xxi.

8. *Op. Cit.*, Turner, p. 219.

9. Brown, G. Burniston, The listener, 26 July 1973 (under LETTERS). See also, *American Journal of Physics*, Vol. 44, No. 8, Aug. 1976, p. 801.

10. Joos, George, Theoretical physics, third edition. Blackie & Sons, 1964, p. 258.

11. *Ibid.*

12. Harwit, Martin, Astrophysical concepts. John Wiley & Sons, 1973, p. 178.

CHAPTER 3
Electric Theory of Inertial Mass

3-1 What is mass?

In physics the four most basic quantities are *mass, length, time,* and *electric charge.* It is not easy to define these basic quantities but we seem to know what these quantities are. We can easily measure these quantities as some multiple of a standard unit of the quantity. Distance can be measured in terms of the standard unit of length, the *meter.* Time can be measured in terms of the standard unit of time, the *second.* Mass can be measured in terms of the standard unit of mass, the *kilogram.* Electric charge can be measured in terms of the standard unit of charge, the *coulomb.*

We do not ordinarily try to define the quantity mass, but we do define the *unit* of mass, the kilogram. The kilogram is the mass of a standard platinum bar that is kept in France as the international standard unit of mass. Mass is measured in terms of that standard kilogram. A very basic instrument for measuring mass is the *analytical balance* found in chemistry and physics laboratories. The mass to be measured is put on one side and known amounts of mass are placed on the other side until there is a balance.

One word of caution. Do not get the quantities *mass* and *weight* mixed up! Never tell a person to lose weight. Tell him to lose mass. "Fat" is associated with mass. Weight is the pull of gravity. It is possible for a person to lose weight by going to the moon. If his "fat" is still there he has not lost any mass.

Ordinarily quantities in physics are defined by equations.

For example, *force* can be defined by Newton's second law which we write as

$$F = ma$$

meaning that *force is equal to mass times acceleration*. This equation tells us that mass tends to "resist" being accelerated. It takes more force to cause a massive body to be accelerated a certain amount than to accelerate a less massive body that same amount. That "sluggish" property of mass is called *inertia*. We shall refer to this property as the *inertial mass*.

In this chapter we shall show that this sluggishness, this *inertial reaction* to being accelerated, is due to electromagnetic feedback generated the instant an electron or proton is accelerated. It involves the production of a magnetic field and a related induction of an electric field that acts backward on the electron or proton. This phenomenon provides some physical insight into the quantity m, inertial mass, in Newton's second law.

3-2   Charge motion produces a magnetic field

Hans Christian Oersted, a Danish school teacher, was the first to discover that an electric current produced a magnetic field. Up until that time it was thought that there was no connection between magnetism and electricity. Now it is common knowledge that electric current is the motion of electric charges and the motion of an electric charge always produces a magnetic field.

A useful equation for finding the magnetic field H produced by the velocity v of the electric charge is

$$\mathbf{H} = \mathbf{v} \times \mathbf{D} \tag{3-1}$$

This is a vector equation, which reads "**H** is equal to **v** cross **D**." **D** is an electric field produced by the charge. The reason for using a vector equation is that the equation not only gives the *value* of the magnetic field but it also gives the *direction* of that field. Another way of describing the production of a magnetic field is as follows: *The magnetic field* **H** *is equal to the rate of drift of* **D** *lines past that field point.*

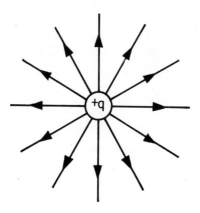

Fig. 3-1 D lines represent the **D** field of the electric charge q. The closer the line spacing the stronger the field strength at that location in the field.

The D field can be pictured (Fig. 3-1) as D *lines* extending out from the electric charge q that produced the field. If the charge moves with velocity **v** those D lines move along with it, with the same velocity. As those D lines move they produce a magnetic field H in the space through which those D lines move.

Figure 3-2 illustrates the direction of magnetic field vector H, at a point distance r from the center of the charge, as the D lines move with velocity **v**. The *direction* of H is obtained by applying the rule for a cross product. That rule tells us that **H** *has the direction a right-hand screw would move if the screw were turned in the same direction as a rotation of the first vector* (**v**) *toward the second vector* (**D**). One can see that H does have that direction in Fig. 3-2.

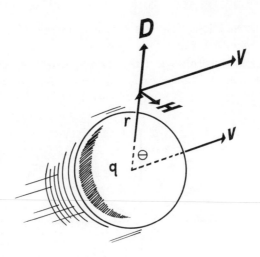

Fig. 3-2 Generation of magnetic field **H** by the motion of D lines associated with charge q as it moves with constant velocity **v**.

The *magnitude* (value) of H in Eq. (3-1) is computed from the equation

$$H = vD \sin \theta \qquad (3\text{-}2)$$

3-3   Inertial mass of an electron

As previously noted the *inertial mass* is mass m in Newton's second law,

$$F = ma \qquad (3\text{-}3)$$

A force applied to an electron will cause it to accelerate. Knowing F and a, one can compute the mass of the electron. We shall deduce an equation for the mass of the electron in terms of its charge q and radius r. During acceleration an electron will move faster and faster. This motion of the electron produces a stronger and stronger magnetic field. We shall show that this changing magnetic field produces an electric field that reacts back on the electron, the inertial reaction.

It can be shown that for nonrelativistic velocities (speed much less than the speed of light) *the kinetic energy is the magnetic energy in the field of the electron.*[1] We shall designate this kinetic/magnetic energy as T. It is a function of v^2 (velocity squared). The power P required to accelerate the electron can be expressed as the derivative of the energy, that is

$$P = \frac{dT}{dt} \tag{3-4}$$

Power can also be expressed as the force times velocity:

$$P = Fv \tag{3-5}$$

Equating these two, one has the equation

$$Fv = \frac{dT}{dt} \tag{3-6}$$

Without really going into the problem of expressing T in terms of charge and distance and v^2, we shall simply note that the calculus operation dT/dt reduces the v^2 to 2 va on the right side of Eq. (3-6). The v divides out of both sides and after including all of the factors the final equation for force turns out to be

$$F = \frac{\mu q^2 a}{6\pi r} \tag{3-7}$$

where μ is the magnetic permeability of free space, q is the charge on the electron, r is the radius of the electron, and a is the acceleration.

In view of Eq. (3-3) (Newton's second law) and Eq. (3-7) the equation for inertial mass of the classical electron at nonrelativistic velocity is

$$m = \frac{\mu q^2}{6\pi r} \tag{3-8}$$

The size of the electron and its charge determine how much inertial mass the electron has.

In Section 8-4 we shall deduce this same equation, Eq. (3-8), for mass of the electron from energy considerations. The mass will be related to the energy in the field of the electron. There we will also deduce the famous equation (see Eq. 8-16)

$$\text{Energy} = mc^2 \tag{3-9}$$

It associates enormous energy with mass m because of the large value of c, the speed of light. However, it appears to the author that the inertia view of mass is very important because it is deduced from Newton's second law, the law that gives us insight into the physical significance of inertial mass. Note that there is a difference between mass and energy; they are not the same. *Mass is the quantitative measure of inertia.* Energy is something else.

Knowing the mass of the electron, we will now compute the radius of the electron from Eq. (3-8). Substituting 9.1096 x 10^{-31} kg as the rest mass of the electron, -1.6022 x 10^{-19} coulomb as the charge on an electron, and 4π x 10^{-7} as the value of μ into Eq. (3-8) yields the value r = 1.8786 x 10^{-15} meter for the *radius of the electron*.

3-4 Inertial mass of a proton

If the preceding derivation of the mass of an electron from classical electromagnetism is valid it should also be valid for the mass of a proton. The assumption is that the proton is a sphere with positive charge q on its surface. Hence the equation for the non-relativistic mass of the proton is also

$$m = \frac{\mu q^2}{6\pi r} \tag{3-10}$$

The value of charge q on the proton is equal to the value of the charge q on the electron. Knowing that the mass of the proton is much greater than the mass of the electron, one can see from Eq. (3-10) that the proton must have a much smaller radius. We shall compute the radius of the proton.

Substituting 1.6726 x 10^{-27} kg as the measured value of rest mass of the proton, +1.6022 x 10^{-19} coulomb as its charge, and $4\pi \times 10^{-7}$ as the value of μ into Eq. (3-10) yields the value r = 1.0232 x 10^{-18} meter for the radius of the proton.

As might be expected these values for the dimensions of the electron and proton are quite different from those proposed by present day modern physicists. Some contend that the electron is a point charge and that the proton has a dimension of approximately 10^{-15} meter. The assumption that the electron is a point charge is nonsense. Such a charge would have infinite electrostatic energy, hardly consistent with the concept of the equivalence of mass and energy.

Furthermore, modern physics has never come up with a satisfactory theory of the electron. Arnold Sommerfeld[2] stated that:

The electron is a stranger in electrodynamics. . .we should face the fact that our electrodynamic theory of the electron is as yet very incomplete.

It is still a real problem in modern physics.

There has been a great deal of experimental work relating to the size of the proton, but it is not a simple matter to measure the size of a proton. Much depends on the reliability of the postulates and theory associated with those measurements. It is one thing to measure the mass of a proton and quite something else to measure its size.

These simple non-relativistic models of the electron and proton are consistent with classical principles. The ratio of the radius of the electron to radius of the proton is 1,836. That is the well known ratio of the mass of the proton to mass of the electron. It is very clear from classical physics that the electrostatic energy in the field of a proton of this small radius would be 1,836 times as large as the electrostatic energy in the field of an electron, confirming the reasonableness of this 1,836 ratio of radii. These relative sizes of the electron and proton make sense from a physical point of view.

3-5   Inertial reaction

Newton's third law refers to a very interesting physical phenomenon to which there is no known exception: *For every action there is an equal and opposite reaction. Action* means a *force* applied to a body by some external source. *Reaction* means the force the body exerts back against the source. The reaction force is instantaneous and precisely equal to the action force.

Newton's second law, $F = ma$, informs us that the applied force F produces an acceleration a on the body of mass m. His third law informs us that the body exerts an equal reaction force F back on the external source. This reaction force is called the *inertial reaction*. The more massive the body is the greater its reaction to being accelerated. That property of reacting against acceleration is called its *inertia*.

If one should kick a brick with his bare toe he would be painfully aware of the reaction force the brick exerts back against his toe. The quantitative value of inertia is mass. The more massive the brick the more its inertial reaction.

We noted in Sections 3-2 and 3-3 that electrons and protons have mass. Any time an electron or a proton is accelerated it exerts an inertial reaction force back on the source. In the next section we shall show that this inertial reaction force is an electric force acting backwards on the charge. When an external source exerts a force on a charge, the charge is accelerated. This acceleration, this changing velocity, generates a changing magnetic field. That changing magnetic field induces an electric field that acts backwards on the charge. That reverse force is the reaction force and it is "felt" back on the source. It makes it more difficult for the external source to accelerate the charge. That phenomenon is the inertial reaction. It is an electric force acting backwards on the electron or proton.

Newton's laws apply to uncharged mass as well as to electric charges. The question may be asked: If the reaction force is an electric force, how could it act upon uncharged mass? That question will be answered later in this book by developing the concept that all ordinary uncharged bodies, including the neutron, are composed of electrons and protons. The position will then be developed that every inertial reaction force is an electric force acting back on all component electric charges in the body.

A complete derivation of the equation for inertial mass in terms of electric charge and acceleration and for the electric mechanism in Newton's third law is given in the author's technical paper published in the March 1983 issue of the Creation Research Society Quarterly.[3]

3-6   Inertial reaction mechanism

The inertial reaction mechanism is an electromagnetic feedback process that takes place any time an electric charge is accelerated. That process may be summarized as follows: Equation (3-1) shows that a magnetic field H is produced when a charge moves with velocity v. Acceleration causes that velocity to increase with time. This changing velocity generates a magnetic field that is *changing* with time. *A changing magnetic field induces an electric field* that exerts a backward force on the charge. This is the inertial reaction, the reaction force in Newton's third law.

To be more specific, we need to introduce Maxwell's induction equation, namely

$$\text{curl } \mathbf{E} = -\dot{\mathbf{B}} \qquad (3\text{-}11)$$

In simple language this great principle tells how electric fields can be generated by a changing magnetic field. The equation says that a *curling* electric field is induced by a changing magnetic field. The electric field is designated by E and the changing magnetic field is denoted by $\dot{\mathbf{B}}$.

The reader need not be concerned here with the switch to B for magnetic field and E for electric field. They are related to the previously employed quantities **H** and **D** by the equations

$$\mathbf{B} = \mu\mathbf{H} \tag{3-12}$$

and

$$\mathbf{D} = \varepsilon\mathbf{E} \tag{3-13}$$

where μ is the magnetic permeability. The magnetic permeability has already been used in Eq. (3-8), the equation for the mass of an electron.

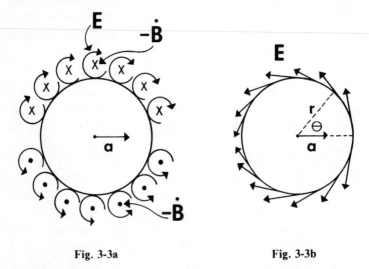

Fig. 3-3a   Fig. 3-3b

Instantaneously induced electric field at the surface of the charge during acceleration. Part (a) shows the directional sense associated with curl **E**= -**Ḃ**. Part (b) shows the electric field **E** at the surface.

The simple physical reasoning presented here is sufficient to illustrate the principles. A more complete treatment is given in reference 3. Figure 3-3a shows the direction of the curling E lines induced around the various positions near the accelerated charge due to the rate of change of magnetic field (-**Ḃ**) in accordance with Maxwell's induction Eq. (3-11).

Since the reaction force takes place *instantaneously* there is not time for any wave propagation. The induced electric field vectors exist only there at the surface of the charge as shown in Fig. 3-3b. Note that all of those E vectors have a rearward component. They all act on that portion of the surface charge in which they are in contact.

The electric force F on a charge q in an electric field E is given by the equation

$$F = qE \tag{3-14}$$

Due to the fact that the electric field vectors E have different values and directions on different points of the spherical charge one has to divide the problem up into little parts (infinitesimals) and integrate (a calculus process of summing up) all of the rearward components of force to get the net electric *reaction force*.

$$F = \frac{\mu q^2 a}{6\pi r} \tag{3-15}$$

Remembering that

$$m = \frac{\mu q^2}{6\pi r}$$

one can see that the electric *reaction* force

$$F = ma \tag{3-16}$$

equals the action force (applied force) $F = ma$ but has the opposite direction.

Hence the mechanism for the inertial reaction is this electromagnetic feedback process at the surface of the accelerated charge. It has nothing directly to do with the energy in the electron.

3-7 Is gravitational mass different from inertial mass?

Technically speaking one should allow for the possibility that there are two different types of mass, *inertial mass* and *gravitational mass*. They are ordinarily considered to be two different manifestations of the same quantity, mass. If this is true, there is no need to use two different names for this quantity.

Whether or not there are two different kinds of mass can be determined, at least in theory, by carrying out with extreme accuracy two types of experiments on the same body. One experiment would be to measure extremely accurately the inertial reaction force on a body and the simultaneous acceleration of that body. The ratio of force to acceleration would, by Newton's second law, give the value of inertial mass. The second experiment would be to somehow measure very accurately the gravitational attraction between that body and a second body of the same dimensions and of precisely the same inertial mass. One could then apply Newton's universal law of gravitation which relates the gravitational force, the distance of separation, and the mass.

If the gravitational experiment yielded a different value of mass from that obtained in the inertial reaction experiment, there would indeed be two different kinds of mass, namely gravitational mass and inertial mass. As of now no one has been able to show experimentally that these are different quantities. Hence, we only use the one word *mass*. If we measure the value of mass on an analytical balance we use the same value for the mass in Newton's three laws of motion and in his universal law of gravitation. That is the way it is actually done in practice. So the practical answer is *no* to the question: Is gravitational mass different from inertial mass? However, the true scientist will always leave open the possibility that some day an experiment may be designed to show that these are two different quantities, not just two different manifestations of the same quantity.

We have seen in this chapter that inertial reaction can be explained in terms of an electric force acting backwards on an accelerated body. We shall develop in the next chapter an electric theory of gravitation in which we explain gravitational attraction as an electric force. Hence, we are attempting to unify the inertial and gravitational properties of mass by showing that they are both electric properties. Up to now we have only dealt with the mass of electric charges but later we shall develop the concept that the mass of all bodies contain electric charges. Hence, electric properties of mass hold equally well for the so-called neutral mass which is not ordinarily thought of as containing electric charges.

References

1. Pemper, Richard R. and Thomas G. Barnes, A new theory of the electron. *Creation Research Society Quarterly*, Vol. 14, March 1978, pp. 216-217. Since that publication Richard R. Pemper has shown in an unpublished treatise that the reason why the magnetic energy is equal to the kinetic energy at non-relativistic speeds is that the increase in the electric field energy with speed is counterbalanced by an almost equal decrease in internal (so-called binding field) energy.
2. Sommerfeld, Arnold, Electrodynamics. Academic Press, 1952, p. 278.
3. Barnes, Thomas G., Electric explanation of inertial mass. *Creation Research Society Quarterly*, Vol. 19, No. 3, March 1983.

CHAPTER 4
Electric Theory of Gravitation

4-1 Newton's universal law of gravitation

According to Newton's universal law of gravitation *every particle of matter in the universe attracts every other particle of matter with a force that is directly proportional to the product of the masses and inversely proportional to the square of their distance apart.* The constant of proportionality is called the universal gravitational constant and is always designated by the capital letter G. It has been evaluated by direct experimental measurement and the value is $G = 6.6732 \times 10^{-11}$ newton meter^2/kilogram^2. When that constant is included, Newton's second law for the gravitational force on particle mass m_1 due to particle mass m_2 at distance r may be written in the form

$$F = \frac{Gm_1m_2}{r^2} = m_1 a \tag{4-1}$$

where acceleration $a = Gm_2/r^2$.

Newton was able to compute the orbits of the planets in the solar system by considering the gravitational force to be the net force on a planet and setting that force equal to mass times acceleration in accordance with his second law of motion. The problem was actually more complicated than implied here. He had to apply the force on every particle of mass in the planet due to every particle of mass in the other bodies. However, he was able to show by the calculus, which he invented to aid him in solving this problem, that if he assumed that the bodies were spheres he could lump the whole mass of each body into Eq. (4-1) and use the distance between the centers of the spheres.

Newton was able to confirm his law of gravitation by checking it with the motion of the planets as described by Johannes Kepler's laws. Kepler's laws of planetary motion were induced from the astronomical observations of Tycho Brahe. We shall show that Newton's gravitational law can be reduced to an electrical phenomenon and expressed as an electrical law.

Some of the great breakthroughs in physics have been achieved when phenomena that were thought to be unrelated have been shown to be closely related. For example, in the early days electricity and magnetism were thought to be unrelated. Later, it was shown that there is always a magnetic field associated with an electric current. Soon afterwards James Clerk Maxwell showed that electricity, magnetism, and light are all electromagnetic phenomena. We shall now attempt to show that gravitation is, in the final analysis, an electric phenomenon.

4-2 Electric basis of gravitational attraction

Knowing the Newtonian equation for the force between two masses, we assume that this gravitational force is an electric force. From that point of view we shall develop a new theory of how an electric field can act on a positive and negative pair of elementary charges causing what would in conventional theory have been a net cancellation of force to be slightly tilted so as to be a net attraction force.[1] The tilting force is so developed as to produce the electric force of the same value as the known gravitational force.

This theory includes a nonlinearity in the field force at the surface of the elementary charge where the self-field of the elementary charge is extremely large, much larger than the breakdown field in any dielectric. The nonlinearity is such that the repulsion electric force is always slightly less than the attraction electric force on the pair of charges, with the balance tilted toward an attraction.

In Sec. 13-4 the neutron is shown to be composed of an electron and proton. It follows that all ordinary uncharged bodies consist of an equal number of positive and negative elementary charges. As a consequence the mass of all these bodies has associated with it equal positive and negative electric field components. We consider these fields to be "dormant" electric components of a gravitational field that cancel everywhere except at the surface of the elementary charges upon which this gravitational field acts. At that surface the aforementioned nonlinearity comes into play.

One might say then that this electric theory of gravitation involves the addition of two new concepts: 1) A nonlinear field effect at the surface of an elementary charge; 2) Mass of all ordinary bodies consists of elementary electric charges. The second concept involves the abandonment of the modern physics concept that light has gravitational mass. Since light has no electric charge it has no mass. This requires a reinterpretation of the "bending" of light rays in the famous solar eclipse experiments. That effect was supposed to be due to the sun's gravitational pull on the star light passing by. Fortunately, scientists such as Charles Poor (See Sec. 6-11) have already shown that there are alternative classical explanations of the star field shifts that have been observed during the times of the famous solar eclipse observations. It is our contention that these eclipse observations, when rightly interpreted, do not support Einstein's general theory of relativity (his theory of gravitation). The time is ripe for a classical type of explanation of gravitation, such as the one we propose.

4-3   Matter's dormant electric fields

We repeat the assumption that all uncharged matter contains an equal number of positive and negative elementary charges. We know that an uncharged atom contains an equal number of electrons and protons. In the author's paper, A

New Theory of the Proton and Neutron, he develops in more detail than that given in Sec. 13-4 the concept that a neutron contains an electron and a proton.[2] This electric composition of a neutron should not be surprising because it is known that a free neutron decays into an electron and proton.

Each of the elementary charges in matter has its own *elementary* electric field. They are independent of each other. Each elementary electric field varies inversely as the square of the distance from its source charge. Since there are an equal number of positive and negative charges in the source matter the net electric field in space is zero. That does not mean that the component positive and negative fields vanish. They are independent of each other. The net field is evaluated by the superposition of those two kinds of vector fields, a process that implies the independence of each of those fields. Figure 4-1 illustrates that superposition process.

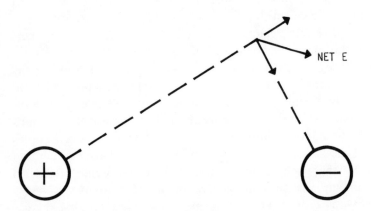

Fig. 4-1 Vector sum of two independent electric fields yields the net electric field.

One might think of these two kinds of electric fields as *partial fields* very much like the *partial pressures* in gases. The law of partial pressures states that the total pressure is the sum of the individual partial pressures. In the case of electric fields the total electric field is the vector sum of the plus and minus electric fields. Those individual electric fields may be thought of as *dormant* fields that have an important part to play in the electric theory of gravitation.

4-4 An overload effect

The gravitation force on an electron or proton is extremely small compared with an ordinary electric force. If the gravitation force is to be an electric force on those elementary charges it must be a very, very small fraction of the ordinary electric force on them. It is, in fact, a second order electric force effect that is herein interpreted as the gravitation force.

Since there are an equal number of protons and electrons in uncharged mass, an electric field acting on that mass would exert an equal attraction and repulsion force on it except for this second order effect. This second order effect always tilts the balance toward an attraction force, the gravitation force. The uncanceling effect is due to nonlinearity in the electric field force at the surface of the proton or electron, yielding a slightly less than expected Coulomb repulsion. As will be shown later, this lessening of the "expected" electric repulsion force is only about one part in 10^{36}. There is no lessening of the expected attraction force. The unbalance, the net attraction, is sufficient to account for the gravitation force.

It should not be surprising that a nonlinearity exists when there is superposition of electric fields at the surface of an electron or proton. The self-field of an electron or proton is extremely large in the region of the charge. It is vastly larger than the breakdown strength of the electric field in air. When a partial field of the same kind (having the same sign) is added to the self-field of an electron or proton, the strength of the field tends to *saturate*. That is to say, it does not quite reach the numeric sum of the two fields. A similar type of field saturation phenomenon is common in electronics. It is called *overloading*.

Arnold Sommerfeld recognized the possibility of non-linearity of electric fields in the region of the electron. He stated it this way:

> *Who can guarantee that the Maxwell equations can be extrapolated right up to the surface or into the interior of the electron? May not their simplicity and linearity be a consequence of the fact that they are exactly valid only for weak fields and that they must be corrected in the immediate neighborhood of concentrated charges. . . in such manner as the theory of dilute solutions in thermochemistry?*[3]

4-5   The reduction factor k

The equation for electric force on an elementary charge q in electric field E is

$$F = qE \qquad (4\text{-}2)$$

Ordinarily, the impressed electric field E_0 is considered to be unaffected by the charge on which it acts. It is now assumed, however, that the *effective* value E of the impressed field is diminished by the *reduction factor* k under certain conditions, so that

$$E = kE_0$$

and

$$F = q(kE_0)$$

The effective value of the impressed field acting on charge q is diminished when the impressed field has the same sign as the charge on which it acts. When a positive field E_0 acts on a proton the effective field is kE_0. When a positive field E_0 acts on an electron there is no overloading. The effective value of the field is E_0, because the field and charge have opposite signs.

An impressed gravitation field is assumed to have both positive and negative partial fields. As previously mentioned they act separately, somewhat like partial pressures in gas. Hence, in a gravitation field it is only the positive partial field that experiences the reduced effect when it acts on a proton. It is only the negative partial field that experiences the reduced effect when it acts on an electron. The sign of the partial field must be the same as that of the charge for the reduction.

When a gravitation field, with its equal positive and negative partial fields, acts on an electron, the result is a net attraction. The attraction of the positive field on the electron is undiminished while the repulsion of the negative field on the electron is lessened. Similarly when a gravitation field acts on a proton the result is a net attraction. The attraction of the negative field on the proton is undiminished while the repulsion of the positive field on the proton is lessened.

4-6   Evaluation of uncanceling factor (1-k)

Let us suppose that the impressed field E_0 is due to a proton at distance r from the elementary charge upon which it acts. The impressed electric field, which is due to the proton with charge q, is given by the well known equation[4]

$$E_0 = \frac{q}{4\pi\varepsilon r^2} \tag{4-3}$$

If that impressed field E_0 were to act on another proton at distance r the repulsive force equation would be

$$F = kE_0 q$$

or in view of Eq. (4-3)

$$F = \frac{kq^2}{4\pi\varepsilon r^2} \tag{4-4}$$

This is slightly less than the attraction force that this impressed field would exert on an electron at distance r. The force of attraction on the electron is given by the equation

$$F = \frac{-q^2}{4\pi\varepsilon r^2} \tag{4-5}$$

Now if we consider the source proton's electric field E_0 acting on a neutron at distance r the net force would be the sum of Eqs. (4-4) and (4-5)

$$F = \frac{-(1-k)q^2}{4\pi\varepsilon r^2} \tag{4-6}$$

because the neutron contains an electron and a proton. This force is the difference in the electric repulsion on the proton part of the neutron and the attraction on the electron part of the neutron. This is the gravitation force a proton would exert on a neutron at distance r from the proton. The uncanceling factor (1-k) causes a net electric attraction force, the gravitation force.

One can evaluate the uncanceling factor (1-k) as follows: write the Newtonian gravitation force on the neutron mass m_n due to proton mass m_p. Equate the gravitation force to the electric force

$$\frac{Gm_p m_n}{r^2} = \frac{(1-k)q^2}{4\pi\varepsilon r^2} \tag{4-7}$$

Solving Eq. (4-7) for (1-k) and using the following values

$$\varepsilon = 8.8542 \times 10^{-12}$$
$$G = 6.6732 \times 10^{-11}$$
$$m_p = 1.6726 \times 10^{-27}$$
$$m_n = 1.6749 \times 10^{-27}$$
$$q = 1.6022 \times 10^{-19}$$

yields

$$(1-k) = 8.103 \times 10^{-37}$$

for the uncanceling factor. This means that the *gravitational force* is only about one part in 10^{36} of what the *electric force* would be on the electric constituents of the neutron if there were not this canceling effect. Gravitation forces are indeed much smaller than ordinary electric forces.

4-7   Neutron to neutron electric attraction

The reduction factor k is related to how much nonlinearity exists in the field at the surface of the elementary charge. The reduction factor is *not* the same for an electron as it is for a proton. The reason for this difference is that the electric self-field is larger at the surface of the proton than it is at the surface of the electron. *The stronger self-field overloads easier.* There is more reduction in k and a larger uncanceling factor for the proton than for the electron.

The uncanceling factor $(1-k_p)$ for the proton is 1,836 times as great as the uncanceling factor $(1-k_e)$ for the electron. That is because the electric self-field at the surface of the proton is 1,836 times as large as that of the electron. This is related to the fact that the proton's mass is 1,836 times that of the electron. The reduction factor in the previous section should have been designated as k_p because it is associated with the *proton* in the neutron. There is a reduction in the repulsion force on that proton but no reduction in the attraction force on the electron.

If the source charge in Sec. 4-6 had been an *electron* instead of a proton, exerting a gravitation force on the neutron, we would have called the uncanceling factor $(1-k_e)$. The gravitation attraction of this electron on the neutron would have been less. The *ratio* of the uncanceling factors of the proton and electron is 1,836, the same as the ratio of their masses.

If a neutron gravitationally attracts another neutron the uncanceling factor is $[(1-k_p)+(1-k_e)]$ because the neutron source of that attraction contains both a proton and an electron. Hence the electric force of attraction between two neutrons is

$$F_e = \frac{[(1-k_p)+(1-k_e)]q^2}{4\pi\varepsilon r^2} \qquad (4\text{-}8)$$

Equating this electric force of attraction to the known Newtonian gravitation force

$$\frac{Gm_n m_n}{r^2} = \frac{[(1-k_p)+(1-k_e)]q^2}{4\pi\varepsilon r^2} \qquad (4\text{-}9)$$

from which one can evaluate more precisely the uncanceling factor, the bracket factor, for the electric force between two neutrons separated distance r.

4-8 Total gravitation force

Most gravitation problems involve masses that contain many molecules each of which is made up of atoms. Remembering that the neutron contains a proton and an electron, the total number N of protons in an atom (including those in the neutrons) is its atomic mass number. That is also the number of electrons in the uncharged atom (including those in the neutrons). Let N_1 be the sum of the protons and neutrons in mass M_1 and N_2 be the sum of the protons and neutrons in mass M_2. The electric form of the total gravitation force on mass M_1 due to mass M_2 is

$$F_e = \frac{N_1 N_2 q^2 [(1-k_p)+(1-k_e)]}{4\pi\varepsilon r^2} \tag{4-10}$$

The two masses are considered to be far apart and small, so r may be considered to be the same for all particle distances from source to object.

The Newtonian force can be written in terms of proton mass m_p and electron mass m_e.

$$F_g = \frac{G N_1 N_2 (m_p + m_e)^2}{r^2} \tag{4-11}$$

One can equate Eqs. (4-10) and (4-11) and solve for the uncanceling factor, the bracket factor, as before.

4-9 Conversion to electric form

In order to convert the problem to an all-electric problem the classical models of the proton and electron will be employed to eliminate the proton and electron mass from Eq. (4-11).

From Eqs. (3-8) and (3-10) we may write

$$m_e = \frac{\mu q^2}{6\pi r_e} \tag{4-12}$$

and

$$m_p = \frac{\mu q^2}{6\pi r_p} \tag{4-13}$$

Substituting those equations into Eq. (4-11), equating that to Eq. (4-10) and solving for the uncanceling factor yields the equation

$$\frac{\mu^2 G N_1 N_2 q^4}{36\pi^2 r^2} \left[\frac{1}{r_e} + \frac{1}{r_p}\right]^2 = \frac{N_1 N_2 q^2 [(1-k_p)+(1-k_e)]}{4\pi\epsilon r^2}$$

which reduces to

$$[(1-k_p)+(1-k_e)] = \frac{\mu G q^2}{9\pi c^2} \left[\frac{r_p + r_e}{r_e r_p}\right]^2 \tag{4-14}$$

where r_e and r_p are the radii of the electron and proton and have the value

$$r_e = 1.87 \times 10^{-15} \text{ meter}$$
$$r_p = 1.023 \times 10^{-18} \text{ meter}$$

Evaluating the right side of Eq. (4-14) yields the following value for the uncanceling factor

$$[(1-k_p)+(1-k_e)] = 8.103 \times 10^{-37}$$

This may be considered to be the conversion factor from ordinary electric repulsion components of the masses to their gravitational attraction.

This derivation is limited to the static or low velocity case and does not take into account the altered configuration of the electron in the neutron and the rest of the atom. However, the difference in this conversion factor and any that can be evaluated from the known masses may serve as a guide to future analysis of the configurations of the electron in various particles in high speed motion.

References

1. Barnes, Thomas G., Harold Slusher, Russell Akridge, and Francisco Ramirez. Electric theory of gravitation. *Creation Research Society Quarterly,* Sept. 1982, 19(2):113-116.
2. Barnes, Thomas G. 1980. New proton and neutron models. *Creation Research Society Quarterly* 17(1):42-47.
3. Sommerfeld, Arnold. 1952. Electrodynamics. Academic Press, New York, p. 276.
4. Barnes, Thomas G. 1977. Foundations of electricity and magnetism, third edition. T.G. Barnes, El Paso, TX, p. 104.

CHAPTER 5
Electric Theory of Rotational Mass

5-1 Moment of Inertia

In rotational motion Newton's second law is written as

$$\tau = I\alpha \tag{5-1}$$

where τ is the torque, α is the angular acceleration, and I is the *moment of inertia*. Moment of inertia is a rotational property of mass. It depends not only on the mass but also on how far the mass is from the axis of rotation. For example, the moment of inertia of a ring of mass m and radius r is

$$I = mr^2 \tag{5-2}$$

An analogy between Newton's second law applied to rotation and his second law applied to translation can be seen by comparing Eq. (5-1) with Newton's second law in the form

$$F = ma$$

Linear acceleration a is caused by the applied force F. Angular acceleration α is caused by the applied torque τ. Newton's third law also holds for rotational motion. *For every applied torque there is an equal and opposite reaction torque.*

We shall show that Newton's laws for rotation can be explained in terms of electromagnetic properties very much like his laws for translation were explained in terms of electrical properties in Chapter 3. The moment of inertia (rotational mass) will be expressed in terms of an electric quantity. This is one more step in our attempt to unify the laws of physics into classical laws of electricity and magnetism.

5-2 Magnetic properties of the electron and proton

Both the electron and proton are considered to have a *spin*. The electron and the proton are assumed to be spheres with charge q uniformly distributed over the surface. The proton's charge is positive and the electron's charge is negative. The spin is the rotational motion of the charge about its axis.

This rotating charge produces a magnetic field. Outside of the charge this has the exact field of an ideal *magnetic dipole*.[1] It is equivalent to a spherical permanent magnet with *uniform magnetization M*. The permanent magnet has an amperian surface current equivalent to the surface current of the rotating surface charge of the spinning electron or proton. When the magnetization is uniform, as in this case, the *magnetic moment* M is equal to the product of the volume times the magnetization. For this sphere the magnetic moment

$$M = \frac{4\pi r^3 M}{3} \tag{5-3}$$

The poles of the magnet set up a demagnetizing force within the magnet. When that is included, the equation for the uniform B field inside of the magnet is,[2]

$$B = \frac{2\mu_0 M}{3} \tag{5-4}$$

In view of Eq. (5-3) the internal B field can be expressed in terms of the magnetic moment.

$$B = \frac{\mu_0 M}{2\pi r^3} \tag{5-5}$$

where μ_0 is the permeability of free space and r is the radius of the sphere.

The magnetic flux ϕ through the cross section of radius

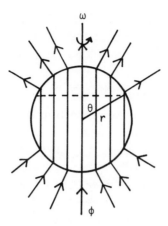

Fig. 5-1 Magnetic flux ϕ of a spinning proton.

r sin θ (see Fig. 5-1) is equal to the product of the cross section area times B

$$\phi = \frac{\mu_0 M \sin^2\theta}{2r} \qquad (5\text{-}6)$$

The magnetic moment may be expressed in terms of angular velocity ω of the rotating spherical charge by the equation

$$M = \frac{q\omega r^2}{3} \qquad (5\text{-}7)$$

That equation is obtained by integrating the elementary magnetic moment dM over the whole sphere. The elementary magnetic moment dM is equal to the current dI times the area enclosed by that ring element. The flux written in terms of the angular velocity is

$$\phi = \frac{\mu_0 q\omega r \sin^2\theta}{6} \qquad (5\text{-}8)$$

5-3   Reaction to angular acceleration

Assume an angular acceleration

$$\frac{d\omega}{dt} = \alpha \qquad (5\text{-}9)$$

of the spherical charge. Differentiating Eq. (5-8) with respect to time yields the rate of change of flux

$$\frac{d\phi}{dt} = \frac{\mu_0 q r \alpha \sin^2\theta}{6} \qquad (5\text{-}10)$$

through each incremental ring.

According to Faraday's induction law when expressed in terms of induced electric field E (instead of emf)

$$2\pi r \sin\theta E = \frac{d\phi}{dt} \qquad (5\text{-}11)$$

This induced electric field is a *reaction field* acting backwards on the moving surface charge. It produces a *reaction torque* because each element of reaction force acts on a lever arm distance r sin θ from the axis. Each element dτ of reaction torque is due to backward force E dq acting on the lever arm. (Fig. 5-2)

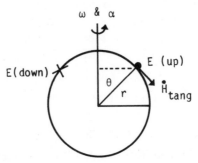

Fig. 5-2 Induced electric reaction of field E at the surface encircles the proton in such a direction as to oppose the angular acceleration α. Tangential component of $\dot{\mathbf{H}}$ near the surface induces this reaction E.

$$d\tau = E \ dq \ r \ \sin\theta \qquad (5\text{-}12)$$

The charge element

$$dq = \frac{q \ \sin\theta \ d\theta}{2} \qquad (5\text{-}13)$$

From Eqs. (5-10), (5-11), (5-12), and (5-13)

$$d\tau = \frac{\mu_0 q^2 r \alpha \sin^3\theta \ d\theta}{24\pi} \tag{5-14}$$

Integrating this over the whole sphere from $\theta = 0$ to $\theta = \pi$

$$\tau = \frac{\mu_0 q^2 r \alpha}{18\pi} \tag{5-15}$$

this is the total *reaction torque*. It is due to the angular acceleration caused by some external torque applied to the spinning charge.

It is obvious from Newton's second law

$$\tau = I\alpha \tag{5-16}$$

and Eq. (5-15) that the moment of inertia

$$I = \frac{\mu_0 q^2 r}{18\pi} \tag{5-17}$$

This shows that the *moment of inertia is an electrical property*. Under the assumption that all uncharged bodies are composed of an equal number of positive and negative elementary particles, the *mass and the moment of inertia of all bodies are electrical properties* and can be expressed in terms of electric charge q.

5-4   Lenz's law confirms inertial reaction

There is a simpler explanation of the inertial torque reaction deduced in Sec. 5-3. Lenz's law gives an explanation of Newton's third law for rotating bodies and does it in terms of electrical quantities. It is quite general because Lenz's law is based on the law of conservation of energy.

Lenz's law tells us that for every rate of change of magnetic flux there is an induced electric field that tends to set up a current that would produce a magnetic flux that opposes that action. If that were not true, there would be a violation of the law of conservation of energy. One would be getting more energy out of the system than he put in. That would be a violation of the law of nature and of common sense.

Note that the magnetic flux in Fig. 5-1 inside the sphere is directed upward. It is due to the electric current resulting from the charge (+ charge) moving in the direction shown. If that rotation is accelerated in the same direction, Lenz's law requires the induced electric field to be in such a direction as to tend to produce a magnetic flux in a direction that opposes the action. That is to say the induced magnetic flux would be downward inside the sphere, opposing the increasing magnetic flux that points upward. That means that the induced E field at the charge on the surface must be directed so as to oppose the acceleration of the rotation.

Hence by this application of Lenz's law one can easily see that the reaction torque deduced in the previous section does have the *reaction* direction and that the moment of inertia is an electrical property as shown in Eq. (5-17).

5-5   Magnetic energy

Now that the moment of inertia has been shown to be an electric property, the conventional mechanical equation for rotational kinetic energy may be used to obtain the *magnetic energy* of a spinning spherical charge. The conventional mechanical equation is

$$\text{Kinetic energy} = \frac{I\omega^2}{2} \tag{5-18}$$

By aid of Eq. (5-17) this may be written as

$$\text{Kinetic energy} = \frac{\mu_0 q^2 r \omega^2}{36\pi} \tag{5-19}$$

This is an electrical equation for the total *magnetic energy* of a spinning spherical charge such as the proton or electron.

Solving Eq. (5-7) for ω and substituting into Eq. (5-19) shows that this magnetic energy is a function of the magnetic moment.

$$\text{Magnetic energy} = \frac{\mu_0 M^2}{4\pi r^3} \qquad (5\text{-}20)$$

This equation can be used to compute the total magnetic energy of an electron or a proton. This equation for magnetic field energy is the same as the equation

$$w = \frac{\mu_0 M^2}{4\pi r^3} \qquad (5\text{-}21)$$

which the author had previously derived, by a different method, for the magnetic energy in a uniformly magnetized spherical permanent magnet.[3] This provides a check on the above derivation of the electrical nature of moment of inertia and on the reduction of the conventional mechanical properties to electrical properties.

5-6 Magnetic mass

The total energy of the proton or electron must include the magnetic energy associated with its spin. In Sec. (8-3) it is shown that the energy associated with a charge in linear motion is mc^2. That mc^2 does not include the magnetic energy of the spinning electron or proton. This magnetic energy due to the spin must be included in their total energy. See Appendix II for the illustration of mass equivalence of the magnetic energy associated with the acceleration of electric current. The spinning electron or proton is equivalent to an electric current.

In the previous section we derived the equation for the magnetic energy of the spinning proton. We were able to express that magnetic energy in several ways: 1) in terms of magnetic moment and radius, 2) in terms of charge and angular velocity of the spin, and 3) in terms of moment of inertia and angular velocity of the spin. Regardless of how one expresses it that intrinsic magnetic energy must be added to the translational mc^2 energy to obtain the total intrinsic energy. Choosing the third way of expressing the magnetic energy, one may express the *total intrinsic energy of the proton or electron* in the following simple, but accurate, form

$$\text{Energy} = mc^2 + \frac{I\omega^2}{2} \qquad (5\text{-}22)$$

where I is the moment of inertia and ω is the angular velocity of spin.

Making use of Eq. (5-20) an alternate, and useful form, of the equation for total intrinsic energy is

$$\text{Energy} = mc^2 + \frac{\mu_0 M^2}{4\pi r^3} \qquad (5\text{-}23)$$

where M is the magnetic moment and r is the radius. In view of the extremely small value of the proton radius, namely 1.0232×10^{-18} meter, the intrinsic magnetic energy is much larger than the translational mass energy mc^2. Using the value $M = 1.4106 \times 10^{-26}$ for the magnetic moment of the proton its intrinsic magnetic energy computed from the second term in Eq. (5-23) yields

$$\text{proton magnetic energy} = 1.857 \times 10^{-5} \text{ joule.}$$

Using the value $m = 1.6726 \times 10^{-27}$ kg for the mass of the proton and $c = 2.9979 \times 10^8$ m/sec for the speed of light, the non-rotational rest mass energy

$$mc^2 = 1.5032 \times 10^{-10} \text{ joule.}$$

This shows that the magnetic energy of the proton is 123,500 times as large as its mc^2 energy in the non-rotational model.

Similarly one may use the same energy equations for the electron. Taking its magnetic moment $M = 9.2849 \times 10^{-24}$ and its radius $r = 1.8786 \times 10^{-15}$ meter, the

$$\text{electron magnetic energy} = 1.3003 \times 10^{-9} \text{ joule.}$$

Taking its mass $m = 9.1096 \times 10^{-31}$ kg, the non-rotational model's rest mass energy

$$mc^2 = 8.1872 \times 10^{-14} \text{ joule.}$$

This shows that the electron's magnetic energy is 15,883 times as large as its non-rotational energy in the non-rotational model used most often in this book.

In Chapter 3 the electric theory of inertial mass was derived from the electric force associated with translational acceleration of *charge.* In view of Appendix II, which shows an inertial reaction force associated with the electric current, as in spin, we assume that there is inertial mass which we call *magnetic mass,* and that it can be computed from the magnetic energy of spin divided by c^2. *Hence we always associate the source of mass with charge or electric current.* Because the intent of this book is to illustrate simple classical approaches, the models have been primarily illustrative of the *charge-related mass.* The magnetic mass models are yet to be developed. They will require among other things *changes in the size* of the electron and proton models.

5-7 Induction vs. radiation

Any time an electric charge is accelerated in translational or rotational motion an electromagnetic wave will be radiated from that source. The amount of radiated energy is negligible in many instances. To have efficient radiation from a radio antenna the frequency has to be high enough for the resultant wavelength of the radiated wave to have dimensions of the order of magnitude of the length of the antenna.

For a single electron to radiate light, the frequency has to be very, very high. In most practical cases it is not just acceleration but also a very high frequency that is needed to produce useful radiation. However, there is some radiation every time a charge is accelerated. Acceleration also produces the induction field energy.

The magnetic field of the spinning spherical charge discussed in the prior sections is called an *induction field* as distinguished from a *radiation field.* In radiation the energy goes on out into space. In induction the energy is propagated out into that field and stays there until deceleration brings it back to the source.

Equation (5-20) is the magnetic energy in the induction field. The radiated energy was considered to be negligible in that derivation. It was assumed, for example, that the magnetic flux ϕ was already developed inside of the sphere. That could not take place instantaneously. Whereas Newton's third law requires the reaction to be instantaneous. It takes some time, albeit an extremely small time, for the wave to fill up the inside of the magnet to produce the demagnetizing effect. However, that delay can usually be neglected.

If the frequency of oscillation (angular acceleration reversing back and forth) were *extremely* high the radiation effects would have to be taken into account. That type of radiation is somewhat like the radiation of radio waves from a small loop antenna. The loop antenna problem has already been solved classically in electrical engineering.[4]

References

1. Jackson, J.D. Electrodynamics, second edition. John Wiley, 1974, p. 195.
2. Barnes, Thomas G. Foundations of electricity and magnetism, third edition, Publisher, Thomas G. Barnes, El Paso, TX, 1977, p. 360.
3. *Op. Cit.*, p. 362.
4. Krauss, John D. Antennas. McGraw-Hill, 1950, pp. 166-167.

CHAPTER 6
Classical Physics -
the Better Physics

6-1 Discordant cosmology

Cosmologists have welcomed the philosophy of modern physics with its release from the common sense constraints of classical physics. Their visionary adaptations of Einstein's relativity have been popular with the news media. *Black Holes* and *Big Bang Theory* are just what the news media like. Unfortunately those theories are not good physics.

The classical theories of light and gravitation proposed in this book are sufficient to "extinguish" the black hole theory. The Big Bang theory is self-contradictory and sheer nonsense. For an excellent refutation of that theory one should read Dr. Harold S. Slusher's *The Origin of the Universe, an Examination of the Big Bang and Steady-State Cosmologies.*[1]

It is not the purpose of this book to go into cosmology, but the following illustration of a very basic problem in astronomy should be pointed out. There is a raging controversy over the interpretation of astronomical redshifts. What makes this particularly important is that Hubble's law, the primary means for determining great distances in astronomy, is in jeopardy.

Dr. Halton Arp has made a credible case against the astronomical redshift data upon which Hubble's law and the expanding universe theory depend. Dr. Arp has shown many cases of *discordant redshifts* that contradict present theory. For some of the details reference is made to a debate between Arp and Dr. John N. Bahcall, both noted scientists. It has been published under the title *The Redshift Controversy*. The following quotes from the moderator of this debate indicates the significance of this challenge to conventional cosmology mounted by Arp.

In the past few years some astronomers have become in- creasingly convinced that there is something basically wrong with the conventional picture of the Universe. They question whether the redshifts of all galaxies are really due to the expansion of the Universe, as has been accepted since the 1920's. They believe that at least some redshifts are dis- cordant, in that they cannot be attributed to the expansion. If they are right, modern cosmology is called into question.[2]

If Dr. Arp's interpretation is correct, a revolution in astronomy, and perhaps in physics, is in the offing.[3]

6-2   Modern physics concepts replaced by classical concepts

We believe that the revolution in physics is long overdue. It is not just this work of Arp. It is the work of a host of other scientists such as Herbert Dingle and Burniston Brown that expose drastic weaknesses in the very foundations of modern physics. The problem is the magnitude of the task of develop- ing a complete alternative to modern physics. In spite of its illogical foundations, the concepts of modern physics are deeply ingrained in the whole warp and woof of physics today. To accomplish the needed mission there must be a complete breakaway from the philosophy and nonsensical concepts of modern physics.

The following concepts and principles of modern physics are rejected in the proposed alternative theories. This restores the concepts to those of classical physics:

1) The modern physics dual nature, the *particle/wave*

nature, is rejected. It is either a particle or a wave, not both. Light is a wave, not a particle. An electron is a particle, not a wave.

2) The principle of equivalent of mass and energy is revised. Mass is one thing; energy is another thing. The sources of inertial mass are charge or electric current. They provide an electric reaction force, inertial effect, when accelerated. Light has energy but no inertial nor gravitational mass because it has no electric charge.

3) As previously mentioned the special theory of relativity concepts of "time dilation" and "space contraction" are rejected. Time and space are assumed to be independent of the state of the "observer."

4) The quantum concept of spin, that spin can only have one fixed value, is rejected. Spin is rotation of charge, an electric current that obeys the law of electromagnetic induction.

5) The modern physics interpretation of the quantum of energy hν is revised. The quantity of energy hν is considered to be the amount of energy absorbed by an atom when it emits a photoelectron. That energy is absorbed from light waves of that frequency, but not from a single photon. The photon "particle" concept is rejected.

6-3 The electric charge concept of a particle

There is *direct evidence* that the electron is a particle. Its velocity can be measured by timing it between two points. Its trajectory can be predicted accurately, much like the trajectory of a bullet. These calculations are employed in the design of the video tube of a television set and in the design of devices such as the betatron. No one questions this evidence of the particle nature of an electron.

The wave nature of particles was first suggested by Louis de Broglie in 1924. He proposed, without proof, the equation

$$\lambda = \frac{h}{mv} \tag{6-1}$$

where λ is the wavelength, h Planck's constant, m the mass, and v the velocity. One author[4] states that this "marked the beginning of quantum theory." Later a famous experiment by Davidson and Germer gave evidence of a Bragg-type diffraction pattern produced by an electron beam impinging on a crystal surface. That experiment does provide a reasonable argument for the wave nature of an electron. However, there is always the possibility that an alternate explanation can be found. One problem is that not many scientists are looking for an alternate explanation.

There are some factors that should be taken into consideration. A classical electron will generate an electromagnetic wave in that experiment even though the electron itself is not a wave. When a beam of electrons impinges on a periodic structure, such as the lattice in a crystal, electromagnetic waves will be generated. They will produce systematic reaction forces that affect the scattering of those electrons.

The author does not intend to undertake the task of making a detailed study and refutation of this experiment or every other experiment that is reputed to support the present status of modern physics. As far as he is concerned that task is somewhat like the task of trying to explain all of the so-called UFO's. He does know the explanation of one or two of them but he does not set a priority on refuting every presumed UFO. The author has set his priority on the task of developing alternative foundations to modern physics that are *consistent, logical, have a cause-and-effect relationship, and have some experimental support.*

Our classification of a particle requires it to have a charge. An electron is a particle. A proton is a particle. A neutron is a particle that is composed of an electron and a proton. Every particle consists of a charge or charges. As in the case of the neutron the charges may add up to zero net charge. If there is no charge there is no particle. Mass is associated with the electric charge. Every particle has mass.

The acceleration of particles causes waves, but waves are not particles. Waves can exert pressure, but that does not mean that the waves consist of a stream of particles. Energy can be propagated by means of waves, but that does not mean that mass is carried along as an ingredient of the wave propagation any more than one would expect a sound wave to be a stream of air. The electromagnetic theory of pressure of light has long been known in classical physics, but that theory has nothing to do with propagation of mass, only the propagation of electromagnetic energy. Electromagnetic waves can be beamed or focused so as to be concentrated on certain regions and vibrate the electrons in a body at that location. So a wave can transport energy.

6-4 A case against the "quantum" of light

There is considerable experimental evidence that light is a wave not a particle. The particle concept of light in modern physics is expressed as a quantum of energy called a *photon*. The quantum of energy is $h\upsilon$ where h is Planck's constant, a very, very small number, and υ is the frequency associated with this particle. We reject that particle concept and hold to the classical theory that light is a wave.

One of the greatest advances in physics was made when James Clerk Maxwell developed the *electromagnetic theory of light*. In that theory light is a wave. There is no photon nor quantum in that theory. Its waves are not governed by the quantum condition $h\upsilon$. It encompasses a wide scope of electromagnetic wave phenomena.

As far back as 1678 Christian Huygens showed that the optics laws of reflection and refraction could be explained on the basis of wave theory. He was also able to explain the optical phenomenon of *double refraction* on the basis of wave theory. Maxwell was able to incorporate all of that into his electromagnetic theory of light.

Direct experimental evidence established the wave nature of light. In 1800 Thomas Young showed that when light from the same source passes through two slits a pattern of interference lines of light will be produced. He computed the wavelength of light in that experiment. Augustin Fresnel did many optics experiments and gave excellent theoretical explanations based on wave theory. The corpuscular theory (particle theory) of light was rejected.

The experimental evidence for the wave theory of light is so strong that one wonders how anyone could reintroduce the particle concept into optics. Possibly it was the tremendous prestige associated with quantum theory that caused experimenters to set their sights solely on confirmation or extension of the quantum theory. It takes a scientist of real courage and originality to challenge the "establishment" in science. Fortunately there are and have been some great ones with that courage and originality. Some have already been mentioned, particularly those who challenged Einstein's relativity theories. There are also those that have challenged quantum theory in one form or another.

One who gave a refutation of the photon of light is the late Herbert E. Ives, a noted scientist at the Bell Telephone Laboratories, whose inventions in optics helped bring about the age of television. In his 1951 Rumford Medal lecture,[5] Ives gave an excellent refutation of the photon of light. His particular target was none other than Einstein's theory of *photoelectric emission.*

He rejects the photon and its associated $h\upsilon$ energy. He accepts $h\upsilon$ as the quantity of energy involved but gives a different interpretation, one that repudiates the particle nature of the incoming light. His insight into this fundamental phenomenon suggests the need for an alternative to the quantum theory of the atom.

Let us consider his case against the photon. Ives' experiments on standing waves in optics and his studies of other experiments led him to reject the photon concept. He gives convincing arguments against the photon. There is no way that he could reconcile small photon particles with the standing wave phenomena in optics, with which he was perhaps the most knowledgeable man in our time.

He was not alone in this rejection of the quantum of light. In his Rumford Medal lecture he gives a lengthy quotation from a 1910 address by H.A. Lorentz. The gist of both Ives' and Lorentz's dissent is as follows: The presumed photons are too small and their waves are incoherent. It would be useless to make a large objective lens for a telescope because the photons would not add in phase, a necessary requirement for greater *definition* in optical devices.

Because of the great importance of this case against the quantum of light the following quote from H.A. Lorentz's 1910 address is included. His arguments are still valid:

Nevertheless the speaker holds the hypothesis of light quanta to be impossible, if the quanta are regarded as wholly incoherent, an assumption which is most natural and which is also made by Planck. The impossibility of incoherent quanta follows from a consideration of interference phenomena. Specifically Lummer and Gehrke have observed interference at a phase difference of two million wavelengths; for yellow light, that corresponds to a length of one meter. If each quantum by itself should be capable of giving sharp interference, then it must also itself extend over that length in the direction of propagation. But, the lateral area of the quantum must also be considerable, which follows from the diffraction theory of optical instruments. Should a light quantum cover only one square centimeter of area, then it would be obviously senseless to fabricate large telescope objectives, for only a small fraction of the area would be used by each quantum in the production of images; on the other hand it is well known that the clearness of images can be greatly enhanced by large objectives. Thus a light quantum would have to be as large as the largest telescope objectives, and, since it is unlikely that the volume of the quantum adjusts itself to our instruments, it would be necessary to conceive of it as considerably larger. But then, only fractions of quanta could get through a smaller opening such as the pupil of the eye. According to the hypothesis of absorption by the retina, only whole quanta can be taken up, so that fractions would have to recombine into whole quanta. Moreover, the consideration of the simplest interference phenomena, such as Newton's rings, shows that the quanta must be divisible, because in the process the ray is resolved by reflection into two parts which travel by different paths and finally come to interference.[6]

6-5 New spin concept

According to quantum theory an electron has a quantum spin, an angular momentum of fixed value $h/4\pi$. It is a vector quantity and may have a plus direction (spin up) or a minus direction (spin down) but can have only this one value. The proton has the same quantum value for its spin and may have either the plus or minus sign.

That quantum concept is rejected and the value of the spin of the electron is assumed not to be limited to a fixed value. The spin of an electron is assumed to be the result of electromagnetic induction as the electron moves in toward the magnetic dipole field of a spinning proton. The rate of change of flux passing through the electron induces a "curling" E field that acts on the electron, causing it to increase its spin. Similarly the changing magnetic flux through the proton increases its spin.

This is a straightforward application of classical electromagnetic theory. The proposed new foundations for modern physics are developed, insofar as possible, to be consistent with the electromagnetic phenomena that are known to work in applied macroscopic applications. It is believed that if the proper models are developed in elementary particle physics and atomic physics that it will fit together and demonstrate the virtue of a return to cause and effect physics.

The proposed models are preliminary and incomplete. Many problems remain to be solved. However, the logic seems to be reasonable. The attempt is to resolve some of the logical contradictions introduced into physics by quantum theory. Lorentz knew that quantum theory does not account for the magnetic energy in the electron. He pointed out that problem to S. Goudsmit and G.E. Uhlenbeck, the scientists who are given credit for the spin concept in quantum theory. Uhlenbeck gives this account of their visit with H.A. Lorentz:

Lorentz received us with his well known great kindness, and he was very much interested, although, I feel, somewhat skeptical too. He promised to think it over. And, in fact, already next week he gave us a manuscript, written in his beautiful handwriting, containing long calculations on the electromagnetic properties of rotating electrons. We could not fully understand it, but it was quite clear that the picture of the rotating electron, if taken seriously, would give rise to serious difficulties. For one thing, the magnetic energy would be so large that by the equivalence of mass and energy the electron would have a larger mass than the proton, or, if one sticks to the known mass, the electron would be bigger than the whole atom! [7]

Goudsmit and Uhlenbeck were also aware of the "relativity problem" for the rotating electron.[8] The rim speed would be faster than the speed of light. We shall show in Sec. 10-11 that there is no problem with the rim speed of a spinning electron exceeding the speed of light, if one accepts the feedback theory developed in Chapter 7.

6-6   Corpuscular and/or wave theory
Even though Sir Isaac Newton did some beautiful experiments with light, experiments that could be explained by wave interference, he held to the corpuscular theory (particle theory). He knew that waves, such as sound waves, bend

around a corner. This is a property called *diffraction*. He saw no evidence of diffraction. So he considered light to be made up of a stream of particles moving in a straight line. He did not realize that the wavelength of light is so short that the diffraction effect is smaller than he had expected.

It took many years for scientists to abandon Newton's corpuscular theory. However, experiments such as Young's double slit experiment and the work of Fresnel and Huygens clearly demonstrated the wave property of light. Great advances were made in optics once the wave nature of light was recognized. It is still the practical theory that is employed in such things as the design of telescopes and contact lenses.

With the advent of quantum theory, light was considered to have a dual nature, a particle nature (photon) and a wave nature. This Dr. Jekyll and Mr. Hyde relationship is acceptable in quantum theory under the Heisenberg indeterminancy principle where one property can be dominant under certain restrictions and the other dominant under other restrictions. The *propagation* of light is ordinarily described in terms of *wave nature*. The *interaction* of light with particles is ordinarily described in terms of the *particle nature*. The laws of conservation of momentum and conservation of energy are then applied to this interaction as they would be to the collision of one billiard ball with another.

The experimental evidence and quantum interpretation of the *photoelectric effect* was the strongest influence in restoring the particle nature of light to physics. Since interference phenomena cannot be explained by particles, the wave property of light must be retained. Otherwise how would one explain light properties of interference and polarization which are straightforward evidences of wave phenomena.

The photoelectric effect is the emission of electrons from the surface of a metal when light strikes the metal. This phenomenon and the peculiar sensitivity of the alkali metals (lithium, sodium, potassium, and particularly rubidium and caesium) has been utilized in the *photoelectric cell* for photo-metry, and has made television possible. The property of particular interest in science has to do with the wave frequency of the incident light and the energy in the emitted electrons. Two important experimental facts seem to be very well established: 1) The incident light must have a frequency high enough (wavelength short enough) for that metal or there will be no photo emission. 2) the *maximum* ejection speed (or energy) with which the electron is emitted depends on the frequency of the incoming light but not on the intensity of the incident light.

The photoelectric emission equation may be written as

$$h\nu = \tfrac{1}{2} mv_{max}^2 + \phi \tag{6-2}$$

where ϕ is the *work function,* the energy required to take the electron out of the surface of the metal. The quantity on the left is the familiar quantum of energy $h\nu$, where h is Planck's constant and ν is the frequency of the incident light.

Einstein introduced the concept that this is the quantum of energy contained in one photon of the incident light. One photon delivers all of its energy to the one electron. He made use of Planck's equation for quantum of energy w, namely

$$w = h\nu \tag{6-3}$$

He received the Nobel prize for this work. This introduction of the quantum of energy is credited as one of the important steps in the development of quantum theory.

Interestingly enough Einstein rejected the later philosophical development of the quantum theory which emphasizes the chance probability at the expense of cause and effect. To emphasize his objection he made the famous statement in German which may be translated as: "I do not believe God Almighty throws dice."

6-7   Photoelectric effect reinterpreted

A case against the photon nature of light was given in Sec. 6-4. Herbert Ives and H.A. Lorentz's analysis of standing waves in optics should be sufficient to lay the particle theory (the photon theory) to rest. No one has yet satisfactorily answered their objections to the photon. Nevertheless, the Einstein interpretation of the photoelectric effect still dominates the science literature.

Herbert Ives' experimental work on photoemission yields the same photoelectric equation as Eq. (6-2) but with no photon involved. It involves standing waves of polarized light, not a quantum corpuscle of energy. He associates the quantum of energy $h\upsilon$ with an internal resonance within the atom, not in the incident light.

Remember that Ives had already shown the photon to be totally inconsistent with standing waves. This experiment employed standing waves. Furthermore it involved two different planes of polarized light and various angles of incidence.

It is well to consider these additional quotes from Ives' famous Rumford Medal lecture:

The significance of these experiments, which I wish to impress upon you, is that it is the *optical* properties of the material, and the optical conditions of the region in which the phenomena occur, which explain these outstanding effects. I stress this because other attempts at explanation, which have been made from quantum mechanical considerations, have almost uniformly neglected or ignored the optical factors. These instead of being secondary or negligible, really dominate.[9]

In reference to the recurrent question of waves vs. corpuscles:

Refuge has been taken from this unsatisfactory state of affairs by using wave descriptions for some phenomena and photon descriptions for others, and it is claimed that the two types of phenomena are never met in the same experiment. I submit that the last experiment. . .certainly comes very close to showing both types of phenomena in conjunction. Electrons are emitted with energy hv, but they follow minutely the orders of the standing wave pattern.[10]

He concludes that it is "extremely difficult if not impossible to retain the idea of light as consisting of discrete photons."[11]

6-8 Where does the quantum hv reside?

The quantum of energy hv is involved in the photoelectric effect as seen in Eq. (6-2). That point is not questioned. That is presumably verified by experiment. The question is whether the quantum of energy resides in the incident light as a photon or in the molecule. It is our contention that the quantum of energy resides in the molecule, not as a separate particle in the incident light wave. We find support in this from some other scientists.

Herbert Ives not only subscribed to that interpretation of Eq. (6-2), but points out that Henri Poincaré suggested it long ago. In reference to his own rejection of the photon Ives states:

To this expression of dissent is to be added that of Henri Poincaré, who in his only published reference to Einstein remarks that while Einstein would put the quantum in the incident light, he, Poincaré, would place it in the molecule; and he speaks elsewhere of finding the "vestibule" by which the molecule admits or releases energy in quanta.[12]

We believe that is the correct interpretation. The electron released from the molecule receives energy hv from the molecule. The frequency v is the frequency of the incident light and energy hv is absorbed from the light waves by the molecule. But this energy extracted from the light was simply a portion of a much greater total amount of energy in the incident light. There is nothing to say that this light is made up of particles of energy hv. We believe that the light is only waves, not particles. It is similar to acoustic waves that impinge upon one's ear. Some of the acoustic energy is absorbed in this wave phenomenon, but no sound particles are involved. Similarly the molecules absorb energy from light waves. No light particles are needed in the process. One should note that Eq. (6-2) relates to only those electrons that are emitted with the *maximum* velocity. Those electrons that are emitted at lower velocities do not fit this quantum equation. They either required more than the work function energy to get out of the surface, or the vestibule of the molecule absorbed less than a quantum of energy. The point to be made is that this sounds very much like a resonant system within the molecule. The quantum constraint is on the *maximum* energy that can be absorbed or radiated per electron. That is all that one can read into Eq. (6-2).

6-9   The Compton effect

Perhaps the strongest support for the photon theory of light is the Compton effect. In 1922 A.H. Compton studied the scattered radiation produced by a beam of monochromatic X-ray impinging on some light elements such as carbon. It was found that the scattered radiation contained not only the frequency present in the incident X-rays but an *additional frequency* produced by the scattering process. The following quote from an elementary physics textbook illustrates the Compton effect arguments put forth in support of the photon.[13]

The mere presence of a new frequency would be most puzzling on the electromagnetic wave theory. Thus in the analogous case of a radio wave, no amount of reflection or scattering by receiving aerials will produce in the wave any frequency not present in the original radiation. Still more remarkable, however, Compton was able to calculate the frequency of the new spectrum line by assuming that it was produced by the collision of a photon of the incident radiation with a free electron of the scattering material. This calculation is particularly interesting as it shows how completely every attribute of radiation except its corpuscular property can be overlooked when dealing with an experiment of this character.

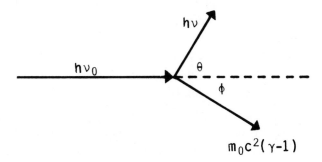

Fig. 6-1 Impact between photon and electron.

[Figure 6-1] represents a photon of frequency v_0 and a consequent energy $h v_0$ colliding with an electron. Photon and electron are imagined at collision to behave exactly as two perfectly elastic billiard balls, the direction and magnitude of their subsequent motion being completely determined by the Laws of the Conservation of Energy and Momentum. The photon is represented as continuing on at an angle θ to its original direction, and the electron at an angle ϕ. Since some of the original energy of the photon has been passed over to the electron, its energy $h v$ after collision is less than that $h v_0$ before collision, its frequency is correspondingly less, and its wave-length greater.

Compton's equation for the observed shift in wavelength is

$$\lambda - \lambda_0 = \frac{h}{m_0 c} (1 - \cos \theta) \tag{6-4}$$

It is to be observed that the difference between the new and the original wave-length is zero for radiation scattered in

the direction of the incident X-rays ($\theta=0$). It becomes, however, progressively greater as θ is increased, until at $\theta=180°$
it is $2h/m_0c$. This characteristic of the phenomenon as well
as the absolute magnitude of the shift has been repeatedly
checked experimentally. Furthermore, a study by C.T.R.
Wilson ray-track method has been successful in identifying
the trails produced by electrons which have shared in this
way the energy of the incident photons.

To emphasize that the photon of light is considered to be *a
particle not a wave* the following quote is taken from a widely
used quantum physics textbook, by Eisberg and Resnick.[14]

Compton (and independently Debye) interpreted his experimental results by postulating that the incoming x-ray
beam was not a wave of frequency ν but a collection of
photons, each of energy $E = h\nu$, and that these photons
collided with free electrons in the scattering target as in a
collision between billiard balls. In this view, the "recoil"
photons emerging from the target make up the scattered
radiation. Since the incident photon transfers some of its
energy to the electron with which it collides, the scattered
photon must have a lower energy E'; it must therefore have
a lower frequency $\nu' = E'/h$, which implies a longer wavelength $\lambda' = c/\nu'$. This point of view accounts qualitatively
for the wavelength shift, $\Delta\lambda=\lambda'-\lambda$. Notice that in the interaction the x-rays are regarded as particles, not as waves,
and that, as distinguished from their behavior in the photoelectric process, the x-ray photons are scattered rather than
absorbed.

There is a real problem with the above postulation "that the
incoming X-ray beam was *not a wave* of frequency ν but a
collection of photons, each of energy $E = h\nu$" and illustrated
by billiard ball collisions. *Those X-rays are electromagnetic
waves that can be polarized. No particle can be polarized like
an electromagnetic wave.* These authors acknowledge (page
45) that X-rays show the "typical transverse wave behavior of

polarization, interference and diffraction that is found in light."

Ralph N. Sansbury[15] suggests a classical explanation of the photoelectric and Compton effects as the production of oscillations of charge" in the scattering material which in turn produce resonant ejection of a photoelectron and/or the secondary X-ray radiation and recoil of a free electron." If this is correct, it is what Poincaré referred to as putting the quantum in the "vestibule" of the molecule.

Sansbury also states:

The mechanism to explain the Compton effect without resort to photons and the need to do so was indicated by Compton himself: "The effect of polarized X-rays on the scattering with change of wavelength has been studied by (Kirchner, F., Annalen der Physik, v. 83, p. 969, 1927). This is an interesting problem because of the difficulty in including polarization phenomena in a purely corpuscular theory of light. . .The result of the experiment indicated that the most probable direction of ejection was perpendicular to the plane of the electric vector (and X-ray beam) in sharp contrast to the behavior of ejected photoelectrons"[Compton and Allison, X-Rays in Theory and Experiment, p. 218, 1950].

As seen in the previous section, Ives has shown that the photon particle concept is untenable. However, he states:

I do not minimize the difficulty of arriving at an explanation of all optical phenomena, such for instance as the Compton effect, exclusively in terms of wave transmission and quantum "vestibules"; nevertheless, on the basis of a long preoccupation with standing waves, I venture to predict that this will be done. . . .[16]

G. Burniston Brown states in his 1982 book that the Compton effect can be treated by wave theory and gives this explanation:

The wave account holds that the electrons are set into oscillation by the X-rays reradiation, and the change in wavelength of some of them is a Doppler effect.[17]

6-10   Distinction between mass and energy

Let us review the case against the equivalence of mass and energy. In Chapter 3 *inertial mass* was shown, by aid of Newton's second law and electromagnetic theory, to be an electric quantity and that it is a measure of *inertia*, the property that "resists" being accelerated. The electric equation for the rest mass of an electron or proton was shown to be

$$m = \frac{\mu q^2}{6\pi r} \qquad (6\text{-}5)$$

As will be seen in Chapter 13 the "neutral" bodies, such as a neutron, have an equal amount of positive and negative charge. The inertial property is associated with both the positive and the negative charges even in neutral bodies.

In Chapter 4 *gravitational mass* was shown, by aid of Newton's law of gravitation, electric theory, and an assumption of asymmetric nonlinearity of the field at the surface of elementary charges in a body, to be an electric property. The question of equivalence of inertial mass and gravitational mass has not been established at the present state of our theoretical development but the values are probably the same.

In Chapter 8 it will be shown that the electrostatic and internal energy in the field of a nonrotating electron or proton is equal to:

$$\text{Energy} = \frac{q^2}{6\pi\varepsilon r} \qquad (6\text{-}6)$$

From Eqs. (6-5) and (6-6) and the equation for speed of light

$c = \dfrac{1}{\sqrt{\mu\varepsilon}}$ one can derive the Einstein equation:

$$\text{Energy} = mc^2 \qquad (6\text{-}7)$$

However, the spin energy was not included in that derivation. It simply shows an alternate way of getting the Einstein equation, whereas each have a spin and associated magnetic energy. That must be included in the total energy. In Chapter 5 it was shown that the magnetic energy of an electron or proton due to its spin is

$$\text{magnetic energy} = \frac{\mu_0 M^2}{4\pi r^3} \tag{6-8}$$

In Sec. 5-6 it was shown that the ratio of magnetic to the non-rotational energy was about 120,000 to 1 for the proton and about 16,000 to 1 for the electron, for the model under consideration.

In view of the relation between magnetic mass and energy, in this model the electron would have about 16,000 times more mass than its measured value of mass. Its gravitational weight would be about 16,000 times greater, if the magnetic energy were included in the equivalence of mass and energy. Similarly, the proton's mass would be about 120,000 times greater than its measured value if this magnetic energy of the spinning proton were included as an equivalent mass.

Although there is this mc^2 energy relationship between the electric-charge mass and also the magnetic (electric-current) mass, one must realize that mass and energy are *different* quantities. Light has energy but does not have mass because mass is associated with electric charge. There is no charge in light. There is no gravitational pull on the energy in light. It has no gravitational mass. The presumed gravitational bending of the starlight by the sun's pull has not been confirmed. In Sec. 6-11 it will be seen by Charles Lane Poor's analysis that the solar eclipse observations have not provided the proof Einstein claimed for it.

6-11   Evidence against Einstein's general theory of relativity

Einstein's second theory of relativity is called *general theory of relativity*. It is a theory of *gravitation*. His first public acclaim came from the presumed confirmation of this theory by British astronomers' observation of deflection of starlight in the vicinity of the sun during a May 29, 1919 solar eclipse. This was one of Einstein's three astronomical "proofs" of general theory. The others were advance of the perihelion of Mercury and gravitational red shift.

All three of these claimed proofs of general theory of relativity are known to be either inconclusive or explainable by classical physics. The advance of the perihelion of Mercury has been shown to be a classical physics phenomenon related to the oblateness of the sun. The best treatment of that effect is a 1982 Master of Science Thesis by Francisco Ramirez.[18] All three of the Einstein astronomical effects have been deduced without relativity. See for example, L. Rongved's 1966 paper "Mechanics in Euclidean terms giving all three Einstein effects."[19]

Even though Einstein's relativity theories, both special theory and general theory, are foundational postulates in modern physics and cosmology there is ample evidence today to reject them both. In fact, evidence against these theories has been present ever since the theories were first published. In his 1922 book, *Gravity versus Relativity,* Charles Lane Poor gave a devastating refutation of the 1919 solar eclipse observational "confirmation" of general theory. He made an extensive study of the photographic plates, the instrumentation, the data, the analysis and the astronomer's report. He shows very objectively that there is no justification for Einstein's claim that these observations confirm his theory. The following three quotes are from his book (p. 197):[20]

This proof as shown in Chapter II rests upon the deflection of light observed by the British astronomers at the total solar eclipse of May 29, 1919. As presented in the 'Report' the evidence makes a strong prima facie case for the Relativity Theory. It was this evidence and the way it was presented to the Royal Astronomical Society at the meeting of November 6, 1919 that caused the furor in regard to the

Einstein theory and the acceptance by so many scientists. But on an examination it will be seen that the strength of this evidence has been greatly magnified, and that *many elements which tend to weaken its force, have been omitted from the public announcements and from popular, or semi-scientific expositions* (emphasis added).

After giving in great detail the setup of the equipment and the technical difficulties and causes of extraneous effects and of the actual data and its interpretation, Poor comes to the following conclusions (pp. 225-226):

When the deflections of light, as actually observed are considered both in direction and in amount, the discordances with the predicted effect become marked, and the plates present little or no evidence to support the relativity theory. Furthermore, if these deflections are real, and not due to instrumental errors (so readily called upon by the relativist to explain everything that the relativity theory cannot account for), then it has not yet been shown that the relativity theory is the only possible explanation. As a matter of fact, there are other perfectly possible explanations of a deflection of a ray of light, explanations based upon every-day, common-place grounds. Abnormal refraction in the earth's atmosphere is one; refraction in the solar envelope is another. The atmospheric conditions under which the eclipse plates were taken were necessarily abnormal, and the plates, themselves, clearly show that the rays of light passed through a mass of matter in the vicinity of the sun; a mass of density sufficient to clearly imprint its picture upon the photographic plates.

Such is the evidence, such are the observations, which, according to Einstein, *"confirm the theory in a thoroughly satisfactory manner."*

Under the chapter heading "The Observed Phenomena and Classical Methods," Poor contends that he has shown the Relativity Theory to be inadequate and that classical methods are adequate (p. 227):

The former chapters show clearly that the Relativity Theory is inadequate to explain either the observed motions of the planets, or the observed deflections of light rays; it can account for the motion of the perihelion of Mercury and for a certain definite deflection of light, but it cannot account for other observed motions of Mercury and Venus, nor for the light deflections as actually observed. The relativist is forced, either to deny the existence of these other motions, or to supplement his theory by some other agency to account for those things, which the relativity theory by itself cannot explain. In the words of the mathematician, the relativity theory, alone, is not sufficient.

On the other hand, the ordinary classical methods of physical and astronomical research can fully explain all the observed phenomena. . .

Eight years later (1930) Poor published in the Journal of the Optical Society of America another devastating criticism of Einstein's claims that the effect of gravitation on light at eclipses had been proved.[21]

There is no valid reason to ignore Charles L. Poor's objective critique. It refutes general relativity. Very few scientists are even aware of Poor's refutation of general theory of relativity. Ignorance and/or bias have been a factor in sustaining this basic tenet of modern physics and cosmology. But there is the need for an alternative. It will be given in Chapter 9.

References

1. Slusher, Harold S. The origin of the universe, an examination of the big bang and steady-state cosmologies. Technical Monograph, Institute for Creation Research, 1982.
2. Field, George B., Halton Arp, and John N. Bahcall. The redshift controversy. W.A. Benjamin Inc., 1973, pp. 3-4.
3. *Op. Cit.* p. 5.

4. McGervey, John D. Introduction to modern physics. Academic Press, 1971, p. 103.
5. Ives, Herbert E. Adventures with standing light waves. *Proceedings of the American Academy of Arts and Science,* Vol. 81, No. 1, pp. 1-31; and reproduced in: Turner, Dean and Richard Hazelett, The Einstein myth and the Ives papers. Devin-Adair, 1979, pp. 194-219.
6. *Op. Cit.,* Ives, pp. 28-29; and Turner pp. 216-217.
7. Jammery, Max. The conceptual development of quantum mechanics. McGraw-Hill, 1966.
8. Eisberg, Robert and Robert Resnick. Quantum physics of atoms, molecules, solids, nuclei, and particles. John Wiley, 1974, p. 301.
9. *Op. Cit.,* Ives pp. 33-34; and Turner pp. 211-212.
10. *Op. Cit.,* Ives pp. 29-30; and Turner pp. 217-218.
11. *Op. Cit.,* Ives p. 31; and Turner p. 219.
12. *Op. Cit.,* Ives p. 29; and Turner p. 217.
13. Miller, Carl W. An introduction to physical science. John Wiley 1932, pp. 349-351.
14. *Op. Cit.,* Eisberg p. 40.
15. Sansbury, Ralph N. Electron structures. *Journal of Classical Physics,* Vol. 1, Part 1, Jan. 1982, pp. 52-53.
16. *Op. Cit.,* Ives p. 31; and Turner p. 219.
17. Brown, G. Burniston. Retarded action-at-a-distance, the change of force with motion. Cortney Publications Luton, Luton Bedfordshire, Great Britain, 1982, p. 72.
18. Ramirez, Francisco. Secular variation on the orbital motion of mercury. Master of Science Thesis, Univ. of Texas at El Paso, Aug. 1982.
19. Rongved, L. 1966 Mechanics in Euclidian terms giving all three Einstein effects, *Il Nuovo Cimento,* XLIV B(2): 255-371.
20. Poor, Charles Lane. 1922. Gravity versus relativity. G.P. Putnam, New York, pp. 197, 225-226, 227.
21. Poor, Charles Lane. *Jour. Opt. Soc. Amer.,* Vol. 20 p. 173, 1930.

CHAPTER 7
Alternative to Einstein's Special Theory of Relativity

7-1 A new theory of electrodynamics

As previously mentioned, in the special theory of relativity *time* and *length are not absolute but depend on relative velocity. Time t^1 is supposed to run slower and space length x^1 to contract in the "moving" frame,* as compared with time t and space length x in the "fixed" frame. This change in time rate is called time *dilation*. Herbert Dingle was correct when he pointed out that *time dilation* leads to a *logical contradiction,* requiring a clock to run both slow and fast at the same time.[1]

Parenthetically it should be pointed out that it is possible that a moving clock might run slow for some physical reason and a moving rod might shorten for some physical reason even though time and space length are not altered by motion. In other words a clock and a meter stick might be made inaccurate by high speed motion. As will be pointed out elsewhere in the book, some of the supposed "proofs" of special theory of relativity are based on experiments which have that problem.

Harold Armstrong has pointed out that a pendulum clock flown around the world would run at an altered rate because the centrifugal force alters the effective pull of gravity.[2] That would obviously change the pendulum clock rate, but that is not time dilation! A high speed muon's decay rate might be altered by a physical process that acts on the particle, a process that will be explained in the new theory developed in this chapter. (See also Sec. 12-9).

The paradoxes (and there are several) in special relativity stem from the abandonment of absolute time and length. A *new theory of electrodynamics, which eliminates the need for special theory of relativity* and eliminates those paradoxes, has been developed by Richard R. Pemper and the author[3, 4, 5]. It restores ordinary time and ordinary length and the associated Euclidean geometry to electrodynamics. It yields the same electric and magnetic field of a uniformly moving charge as the field that has been measured in the laboratory. It should be added that special theory of relativity yields the same field in the laboratory frame of reference. However, these two different theories yield different results when "observed" from different frames of reference.

We shall give an abbreviated development of this new theory of electrodynamics in the sections that follow.[6] One might call this a feedback theory of electrodynamics. A novel part of the theory is the new concept of a feedback between different frames of reference, the laboratory frame and the frame moving with the charge.

7-2   Field of a moving charge

In special theory of relativity the concept of space *length contraction* is employed to derive the field of a moving charge. The contraction takes place in the direction of motion squeezing together the lines of force as in Fig. 7-1b as compared to the static case in Fig. 7-1a.

By means of the *new theory of electrodynamics* one can derive the field equation without resorting to length contraction or time dilation. Length and time are assumed to be ordinary length and ordinary time and the Galilean transformations are assumed to hold.

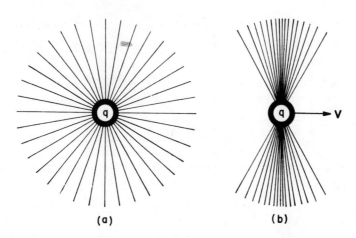

Fig. 7-1a E field for a charge at rest.

Fig. 7-1b E field for a charge in uniform motion with the ratio $v/c = 0.94$.

Making use of the relations

$$\mathbf{D} = \varepsilon\mathbf{E}; \quad \mathbf{H} = \mathbf{v}{\times}\mathbf{D}; \quad \mathbf{B} = \mu\mathbf{H}; \quad \text{and} \quad c = \frac{1}{\sqrt{\mu\varepsilon}}$$

one notes that movement of charge q and the associated E lines *induces* a B field in the S frame (fixed frame) in accordance with

$$\mathbf{B} = \frac{\mathbf{v}{\times}\mathbf{E}}{c^2} \tag{7-1}$$

This movement of E lines (moving with the S^{I} frame) lays down a *track* of B in the S frame.

Substituting the electric intensity E for the charge at rest, namely

$$\mathbf{E} = \frac{q}{4\pi\varepsilon r^2}\,\mathbf{u}_r \tag{7-2}$$

into Eq. (7-1) yields

$$\mathbf{B} = \frac{qv\,\sin\,\theta}{4\pi\varepsilon c^2 r^2}\,\mathbf{u}_\phi \tag{7-3}$$

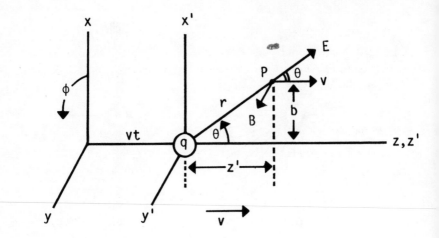

Fig. 7-2 Elementary charge q in uniform motion.

where u_ϕ is the azimuthal unit vector as shown in Fig. 7-2.

The charge is located at the origin of the S^I frame which moves in the z direction with velocity **v**. Since $\sin \theta = b/r$

$$\mathbf{B} = \frac{qvb}{4\pi\varepsilon c^2 r^3} \mathbf{u}_\phi \qquad (7\text{-}4)$$

Since charge q is moving with respect to point P, which is fixed in the S frame, it is obvious that the distance r changes with time. Note that

$$r^2 = b^2 + z'^2 \qquad (7\text{-}5)$$

where b is a constant. Applying the Galilean transformation

$$z' = z - vt \qquad (7\text{-}6)$$

then Eq. (7-5) becomes

$$r^2 = b^2 + (z-vt)^2 \qquad (7\text{-}7)$$

Substituting into Eq. (7-4)

$$\mathbf{B} = \frac{qvb}{4\pi\varepsilon c^2 [b^2 + (z-vt)^2]^{3/2}} \mathbf{u}_\phi \qquad (7\text{-}8)$$

Equation (7-8) shows that B is a variable function of time, increasing as q moves nearer to P and decreasing as q moves away from P. At every point in the fixed frame there exists a $\dot{\mathbf{B}}$. In view of Maxwell's equation

$$\nabla \times \mathbf{E} = -\dot{\mathbf{B}} \tag{7-9}$$

there will be an induced E field that "curls" around the B.

Taking the partial derivative of **B** with respect to time

$$\dot{\mathbf{B}} = \frac{3qv^2b(z-vt)}{4\pi\varepsilon c^2[b^2 + (z-vt)^2]^{5/2}} \mathbf{u}_\phi \tag{7-10}$$

Utilizing Eq. (7-7) this becomes

$$\dot{\mathbf{B}} = \frac{3qv^2b(z-vt)}{4\pi\varepsilon c^2 r^5} \mathbf{u}_\phi$$

Since $\beta = \dfrac{v}{c}$, $\sin\theta = \dfrac{b}{r}$, and $\cos\theta = \dfrac{z-vt}{r}$

one obtains

$$\dot{\mathbf{B}} = \frac{3q\beta^2 \sin\theta \cos\theta}{4\pi\varepsilon r^3} \mathbf{u}_\phi \tag{7-11}$$

Making the substitution

$$E_0 = \frac{q}{4\pi\varepsilon r^2}$$

$$\tag{7-12}$$

yields

$$\dot{\mathbf{B}} = \frac{3\beta^2 E_0 \sin\theta \cos\theta}{r} \mathbf{u}_\phi \tag{7-13}$$

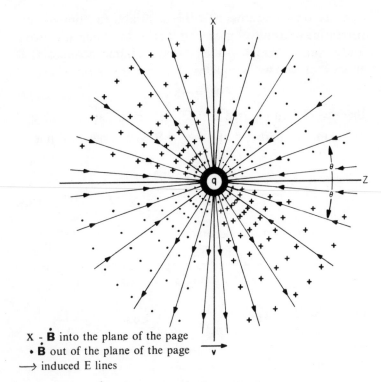

X - $\dot{\mathbf{B}}$ into the plane of the page
• $\dot{\mathbf{B}}$ out of the plane of the page
⟶ induced E lines

Fig. 7-3 Plot of $\dot{\mathbf{B}}$ and the associated induced $\mathbf{E}_f$ lines.

Fig. 7-3 is a plot of the $\dot{\mathbf{B}}$ field given by Eq. (7-13) and the **E** field induced by Maxwell's Eq. (7-9).

7-3 The feedback postulate

Fig. 7-3 shows the E_f field induced in the S frame due to the $\dot{\mathbf{B}}$ in the S frame. *A basic postulate of this new theory is that the induced E_f in the S frame is fed back point-by-point into the S^1 frame, superimposing this induced E_f upon the original field in the moving frame.* This feedback may be expressed mathematically as

$$E = E_0 + E_f \qquad (7-14)$$

where E_0 is the original field as in Eq. (7-12) and E_f is the resultant of the feedback electric field. Feedback alters the original field and there is another and another round (or stage) of feedback. There are an infinite number of stages of feedback and Eq. (7-14) may be written as

$$E = E_0 + E_{f1} + E_{f2} + \cdot \cdot \cdot + E_{fn} + \cdot \cdot \cdot \quad (7\text{-}15)$$

or

$$E = E_0 + \sum_{n=1}^{\infty} E_{fn} \quad (7\text{-}16)$$

Fig. 7-3 shows the E_{f1} field as the induced field. When it is superimposed on the original E_0 field the resultant electric field is diminished in the $+$ and -z directions (parallel to the motion) and increased in the transverse directions. The additional stages, e.g. E_{f2}, E_{f3}, $+ \ldots$ carry that process further, diminishing $E_{\parallel}$ and increasing $E_{\perp}$. The feedback is considered to be *instantaneous, point-by-point* from S to S^1 frame.

7-4 The feedback solution

The radial symmetry in Eq. (7-13) and shown in Fig. 7-3 leads to the postulate that the induced E lines are radial. Under that assumption one may show, by applying Eq. (7-9) to an infinitesimal loop of area $rd\theta dr$, that at every point in the field

$$\frac{dE_f}{d\theta} = r\dot{B} \quad (7\text{-}17)$$

For the first stage of feedback, from Eqs. (7-13) and (7-17) one can write

$$\frac{dE_{f1}}{d\theta} = 3\beta^2 E_0 \sin\theta \cos\theta$$

or

$$\int dE_{f1} = 3\beta^2 E_0 \int_0^\theta \sin\theta \cos\theta \, d\theta$$

Integration over limits $_0$ to θ yields

$$E_{f1}(\theta) - E_{f1}(0) = \frac{3}{2} \beta^2 \sin^2\theta \, E_0 \quad (7\text{-}18)$$

Equation (7-18) gives the *difference* in E_{f1} at θ and $_0$, not the absolute values at θ and $_0$. During feedback the *reference* E_{f1} ($_0$) diminishes; Fig. 7-3 shows $E_{f1}(_0)$ pointing *inward*. That reference value, which is needed to establish the absolute value of $E_{f1}(\theta)$, may be written in the form

$$E_{f1}(0) = -\lambda_1 E_0 \tag{7-19}$$

where λ_1 is the *diminishing factor* in the first stage of feedback. Putting that into Eq. (7-18)

$$E_{f1} = \frac{3}{2} \beta^2 \sin^2\theta \, E_0 - \lambda_1 E_0 \tag{7-20}$$

Note that this is the second term in the series

$$E = E_0 + E_{f1} + E_{f2} + \cdots + E_{fn} + \cdots \tag{7-21}$$

In computing E_{f2}, the second stage of feedback, one begins with

$$B_2 = \frac{\mathbf{v} \times \mathbf{E}_{f1}}{c^2} \tag{7-22}$$

and goes through a similar process to obtain

$$E_{f2} = \frac{15}{8} \beta^4 \sin^4\theta \, E_0 - \lambda_1 \left[\frac{3}{2} \beta^2 \sin^2\theta \, E_0\right] - \lambda_2 E_0 \tag{7-23}$$

After developing the whole sequence of E_{fn} terms and substituting into Eq. (7-21) it can be shown that the series reduces to the closed form

$$E = E_0 \frac{1-\lambda}{[1-\beta^2\sin^2\theta]^{3/2}} \tag{7-24}$$

where λ is the sum of the diminishing factors for each stage of feedback, i.e.

$$\lambda = \sum_{n=1}^{\infty} \lambda_n$$

Equation (7-24) gives the resultant electric field. The task remains of solving for λ. This is done by *postulating the conservation of electric flux* and expressing this mathematically as

$$\iint \mathbf{E \cdot n} dS = \frac{q}{\epsilon} \tag{7-25}$$

Making use of Eqs. (7-24) and (7-12) and employing the Gaussian surface element

$$dS = 2\pi r^2 \sin\theta \ d\theta$$

enables one to write Eq. (7-25) in the form

$$\frac{q(1-\lambda)}{2} \int_0^\pi \frac{\sin\theta \ d\theta}{[1-\beta^2\sin^2\theta]^{3/2}} = q \tag{7-26}$$

Since

$$\int_0^\pi \frac{\sin\theta \ d\theta}{[1-\beta^2\sin^2\theta]^{3/2}} = \frac{2}{1-\beta^2} \tag{7-27}$$

it is obvious that the diminishing factor $\lambda = \beta^2$. Hence Eq. (7-24) becomes

$$E = \frac{E_0 (1-\beta^2)}{[1-\beta^2\sin^2\theta]^{3/2}} \tag{7-28}$$

or when expressed in terms of γ and put in vector form

$$\mathbf{E} = \frac{E_0}{\gamma^2[1-\beta^2\sin^2\theta]^{3/2}} \mathbf{u}_r \tag{7-29}$$

This is the electric field of an elementary charge as modified by the feedback when the charge is moving with constant velocity. In the S frame it is identical with the field that one would obtain from relativistic electrodynamics. Neither length contraction nor time dilation is involved in the derivation of Eq. (7-29) by this new theory of electrodynamics.

In this new theory the E *field is the same in both the S frame and the S' frame.* This follows from the fact that there is neither length contraction nor time dilation. The same E field pattern is "seen" in both frames of reference. The field pattern which is obtained from Eq. (7-29) is shown in Fig. 7-1 for (a) an electron at rest and (b) an electron in motion with a speed of 0.94c or $\gamma = 3$. It is to be noted that for high speed the E lines of flux shift toward the transverse direction. This means a reduced value of E in the direction of motion and a larger value in the transverse direction.

Applying Eq. (7-1) to the solution for E one obtains the solution for B, namely

$$\mathbf{B} = \frac{E_0 \ v \ \sin \ \theta}{c^2 \gamma^2 [1 - \beta^2 \sin^2 \theta]^{3/2}} \ \mathbf{u}_\phi \qquad (7\text{-}30)$$

7-5   Preferred frame of reference

The conclusions drawn from the Michelson-Morley experiment (Sec. 2-4) and other experiments, have led most scientists to reject the concept of an *absolute* frame of reference. However, the successful application of the *feedback principle* to the derivation of the electric and magnetic field of a moving charge (Sec. 7-4) leads one to look for some type of *preferred frame of reference*.

In Sec. 7-4 the preferred frame is the S frame. The $\dot{\mathbf{B}}$ in that frame, *and that frame only*, produces the E_f which is fed back into the moving frame (the S^1 frame). One cannot alter the physical result of this feedback by some "thought" experiment, such as a mathematical set of axes moving with another velocity. The feedback is due to some physical property in the space associated with the S frame. The cause of that property in the space associated with the preferred frame has not yet been established. For the time being one might consider the laboratory to be the preferred frame of reference, a *local ether* concept.

This physical property of the space in the preferred frame of reference will now be employed to resolve a paradox that exists in the special theory of relativity. Consider two elementary positive charges q_1 and q_2 located at the ends of a string and moving with the same velocity **v** with respect to the S

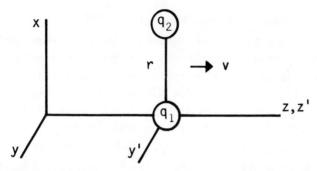

Fig. 7-4 Two charges moving with the same velocity **v**.

frame as shown in Fig. 7-4. The direction of the string is perpendicular to the velocity **v**.

Let S^1 be the moving frame in which the charges are at rest. According to special relativity, the repulsion force exerted on each of these charges in the S^1 frame is the static coulomb force.

$$F^1 = \frac{q_1 q_2}{4\pi\varepsilon r^2} \qquad (7\text{-}31)$$

According to special theory of relativity the repulsion force between these two charges in the S frame is

$$F = \frac{q_1 q_2}{4\pi\varepsilon r^2 \gamma} . \qquad (7\text{-}32)$$

Since r is *perpendicular* to the motion there is no change in r. Hence from special theory of relativity one has the paradoxical situation of the force F^1, Eq. (7-31), yielding a greater force than F, Eq. (7-32). This is an illogical physical situation. For example, a weak string might break under tension force F^1 but not break under tension force F. Physically the string either *would* or *would not* break regardless of the solution method.

The new theory of electrodynamics does not have that paradox. In this theory the force is *equal* in both frames for any value of v. That force turns out to be equal to the special relativity force F in the S frame, Eq. (7-32).

According to the new theory of electrodynamics, when charge q_1 moves with velocity **v** with respect to the preferred frame S it "lays down" a *track* of B in the space associated with the preferred frame. *The B lines do not move along with the charge but reside in the S frame.* Of course the E lines associated with q_1 do move along with the charge.

The charge q_2, also moving along with the S' frame, moves with velocity **v** with respect to q_1's B field in the S frame. Hence, there is a magnetic force on q_2 directed toward q_1. This is a *pinch effect* force.

Similarly, the charge q_2 lays down a track of B in the S frame through which q_1 moves with velocity **v**. The *pinch* force on q_1 is in the direction of q_2.

An *elementary* charge moving with constant velocity *does not "feel" any net reaction force with respect to its own B field,* the field being layed down in the S frame. But it does "feel" the pinch force due to its motion with respect to the B field, in the S frame, of any other charge.

The B field in the S frame, due to the motion of q_1, is given by Eq. (7-30), which includes the feedback effect. The charge q_2 is moving through the B field with velocity **v**. In Eq. (7-30) q_2 is located at distance r and angle $\theta = \pi/2$ with respect to the origin, where q_1 is considered to be. So B at the location of q_2 *always* has the value and direction.

$$\mathbf{B} = \frac{q_1 v \; \gamma}{4\pi\varepsilon c^2 r^2} \mathbf{u}_\phi$$

(7-33)

Since q_2 is moving with velocity v with respect to B, the magnetic pinch force

$$F_{mp} = - \frac{q_1 q_2 \beta^2 \gamma}{4\pi\varepsilon r^2}$$

(7-34)

There is also the electric repulsion force q_2 due to q_1 which from Eq. (7-29) is

$$F_r = \frac{q_1 q_2 \gamma}{4\pi\varepsilon r^2} \qquad (7\text{-}35)$$

The net force on q_2 is

$$F = \frac{q_1 q_2 \gamma}{4\pi\varepsilon r^2} (1 - \beta^2) \qquad (7\text{-}36)$$

but

$$(1 - \beta^2) = \frac{1}{\gamma^2} \qquad (7\text{-}37)$$

so

$$F = \frac{q_1 q_2}{4\pi\varepsilon r^2 \gamma}$$

This force on q_2 *obtained by the new theory equals the relativistic answer* Eq. (7-32) for the S frame. But the relativistic answer Eq. (7-31) for the S^1 frame is different, an untenable problem for relativity. Whereas the *new theory of electrodynamics does not have that problem; it gives the same answer* Eq. (7-37) *in both the S frame and the S^1 frame.*

7-6 Equivalence of forces

There are two reasons why the new theory of electrodynamics yields the same force in the moving (S^1) frame as in the preferred (S) frame: 1) *The same electric field E due to q exists in both frames.* Hence, the *same electric repulsion force*

$$F_r = qE \qquad (7\text{-}38)$$

exists in S^1 as in S. 2) Although the charge *attached* to the moving (S^1) frame *does not see the magnetic pinch force*

$$F_{mp} = q\ v \times B \qquad (7\text{-}39)$$

that exists in the S frame, *it does see an equivalent electric pinch force*

$$F_{ep} = qE_i \qquad (7\text{-}40)$$

The field E_i is the electric field *induced* in the S^1 frame by the relative motion v^1 of the B lines moving *backwards* with respect to the S^1 frame. In view of the Lorentz force

$$E_i = B \times v' \qquad (7\text{-}41)$$

Since $v^1 = -v$ the induced field

$$E_i = v \times B \qquad (7\text{-}42)$$

Substituting this into Eq. (7-40) shows that the solution for the electric pinch force in the S^1 frame is equal to the solution for the magnetic pinch force in the S frame, i.e.

$$F_{ep} = F_{mp}$$

Hence, the electromagnetic force solutions in the moving (S^1) frame *are equivalent* to the electromagnetic force solutions in the preferred (S) frame.

This *equivalence of forces in the two frames of reference* is what one would expect in classical physics. It is an illustration of how the new theory of electrodynamics eliminates special theory of relativity paradoxes and restores classical concepts to electrodynamics.

References

1. Dingle, Herbert. Science at the crossroads. Martin Brian & O'Keefe, London, 1972, p. 45.
2. Barnes, Thomas G., Richard R. Pemper, and Harold L. Armstrong. A classical foundation for electrodynamics, (Reference No. 9) *Creation Research Society Quarterly,* Vol. 14, 1977, p. 45.
3. *Op. Cit.* pp. 38-45.
4. Pemper, Richard R. and Thomas G. Barnes. A classical model of the electron, *Creation Research Society Quarterly* 14(3): 1978, pp. 210-220.
5. Pemper, Richard R. A classical foundation for electrodynamics, Master's Thesis, U.T. El Paso, TX May 1977.
6. Barnes, Thomas G. Foundations of electricity and magnetism, third edition. T.G. Barnes Publisher, 1977, pp. 368-376.

CHAPTER 8
The High Speed Electron

8-1   Classical theory of the electron

Classical theories of the electron were developed by Abraham, Lorentz, and others prior to Einsteinian relativity and quantum theory. They were fairly successful but had some problems that were never completely resolved and did not include a *spin* and the associated magnetic field.

In this chapter we develop a classical theory of the electron that includes the feedback effect at high speed. The electron is deformable, changing shape with speed. It has the following properties*:

1) It has a negative charge of fixed value. This electron's charge is traditionally designated by the letter e, a numeric quantity with a value of approximately 1.6022 x 10^{-19} coulomb. The author departs from tradition and uses the algebraic quantity q for all charges.

2) It has a spherical shape when at rest.

3) At very high velocity it has the shape of an ellipsoid of revolution with the minor axis in the direction of motion.

4) The charge is assumed to reside on the surface.

---

*The spin is neglected in this chapter where the emphasis is on high speed translational mass, not rotational mass.

5) Its electric self-field E exerts an outward tension of $\epsilon E^2/2$ newtons/meter^2 on the surface charge in a direction normal to the surface.

6) There is a nonelectric force inside the electron that keeps it from being pulled apart by the outward electric tension on its surface.

7) Its total energy, nonelectric internal energy plus electrostatic field energy, when at rest is

$$\text{Energy } = \frac{q^2}{6\pi\epsilon r} \tag{8-1}$$

From Eq. (8-1) and $E = mc^2$ (See Sec. 8-6) the rest mass

$$m_0 = \frac{q^2}{6\pi\epsilon c^2 r}$$

By aid of the speed of light equation $c = \dfrac{1}{\sqrt{\mu\epsilon}}$ this rest mass equation may be written as

$$m_0 = \frac{\mu q^2}{6\pi r} \tag{8-2}$$

8) Its *transverse mass* m increases with speed in accordance with the equation

$$m = \frac{m_0}{\sqrt{1-v^2/c^2}} \tag{8-3}$$

Equation (8-3) may also be written as $m = \gamma m_0$ where

$$\gamma = \frac{1}{\sqrt{1-v^2/c^2}}$$

This increase in transverse mass has been deduced from theory and has been measured by experimental means.

9) Its *longitudinal mass* increases in accordance with the equation

$$m = \gamma^3 m_0$$

Our new theory of the electron includes all of the above mentioned properties. It deduces the dynamic ones from the feedback concept developed in the preceding chapter.[1] This provides a physical explanation for those properties.

The distinction between the inertial mass of the electron and its energy has already been brought out in Chapter 3 where inertial mass was associated with electric properties. Similarly, the gravitational mass discussed in Chapter 4 indicated a distinction between mass and energy. We shall follow tradition and assume that inertial mass and gravitational mass are the same, whether it be for an electron, proton, or neutron.

8-2   Electron's electrostatic energy

Assume that an *electron* at rest is a sphere of radius a_0 with total charge q distributed uniformily over its surface. The electric field

$$\mathbf{E} = \frac{q}{4\pi\epsilon r^2} \mathbf{u}_r \tag{8-4}$$

The total *electric field energy*

$$V = \frac{\epsilon}{2} \int E^2 dV \tag{8-5}$$

where use has been made of the electric energy density $\epsilon E^2/2$ joules/meter[3] and the integration is taken over the total volume (V) outside of the electron, from radius a_0 to ∞. In spherical coordinates $dV = 2\pi r^2 \sin\theta \, d\theta \, dr$ so

$$V = \frac{q^2}{16\pi\epsilon} \int_0^\pi \sin\theta \, d\theta \int_{a_0}^\infty \frac{dr}{r^2}$$

which yields

$$V = \frac{q^2}{8\pi\epsilon a_0} \tag{8-6}$$

for the total energy in the electric field of a static electron.

8-3   Rest mass energy of the electron

Recalling that electric lines of force exert a tension stress (force per unit area) $\dfrac{dF}{dA} = \dfrac{1}{2} \epsilon E^2$

and using Eq. (8-4) the electric stress on the electron's surface is

$$\frac{d\mathbf{F}}{dA} = \frac{q^2}{32\pi^2\varepsilon a_0^4} \mathbf{u}_r \tag{8-7}$$

In order for the electron to be in stable equilibrium there must be some internal *binding* force or stress equal and opposite to the outward electric stress. Richard R. Pemper has developed in his Master's Thesis, a model of the electron that meets this requirement.[2] This force is assumed to result from some unconventional type of field continuum distributed throughout the electron's interior. The energy associated with this field is referred to as the *binding field energy* U_{bf} of the electron.

The internal stress is assumed to be

$$d\mathbf{F} = - \frac{q^2 r^2 \sin\theta \ d\theta \ d\phi}{32\pi^2\varepsilon a_0^4} \mathbf{u}_r \tag{8-8}$$

which satisfies the requirement of equilibrium at the surface. The energy dU_{bf} of each binding element is expressed as

$$dU_{bf} = \int_0^{a_0} d\mathbf{F} \cdot d\mathbf{r}$$

and after substituting Eq. (8-8) and integrating

$$dU_{bf} = \frac{q^2 \sin\theta \ d\theta \ d\phi}{96\pi^2\varepsilon a_0}$$

This element is integrated over θ and ϕ to yield the total *binding field energy* in the electron.

$$U_{bf} = \frac{q^2}{24\pi\varepsilon a_0} \tag{8-9}$$

Adding the above binding field energy to the electric field energy V Eq. (8-6) yields the *rest mass energy*

$$U_0 = \frac{q^2}{6\pi\varepsilon a_0} \qquad (8\text{-}10)$$

for the electron. Assuming the values of the rest mass energy as 8.19×10^{-14} joule and the charge as 1.6×10^{-19} coulomb, Eq. (8-10) yields for the radius of the electron

$$a_0 = 1.87 \times 10^{-15} \text{ meters}$$

This is a somewhat smaller value than the so-called classical radius of the electron.

8-4   So-called equivalence of mass and energy

Consider an electron moving with a velocity much less than the speed of light. Its total energy may be expressed as

$$U = U_0 + T \qquad (8\text{-}11)$$

where T is the magnetic field energy resulting from motion of the electron. In this low velocity case the electric field energy and binding energy will remain essentially constant.

The magnetic field energy

$$T = \frac{\mu}{2} \int H^2 dV \qquad (8\text{-}12)$$

where the integration is over the whole volume outside of the electron. Making use of

$$\mathbf{H} = \mathbf{v} \times \mathbf{D} \quad \text{and} \quad \mathbf{D} = \varepsilon\mathbf{E}$$

one may write

$$T = \frac{\mu}{2} \int (v\varepsilon \ E \ \sin\theta)^2 dV$$

noting that $\mu = \dfrac{1}{\varepsilon c^2}$ and $dV = 2\pi r^2 \sin\theta \ d\theta \ dr$

$$T = \frac{\pi v^2 \varepsilon}{c^2} \int_0^\pi \int_{a_0}^\infty E^2 r^2 \sin^3\theta \ d\theta \ dr$$

and substituting for E from Eq. (8-4)

$$T = \frac{q^2 v^2}{16\pi\varepsilon c^2} \int_0^\pi \sin^3\theta \; d\theta \int_{a_0}^\infty \frac{dr}{r^2}$$

which yields

$$T = \frac{1}{2} \left[\frac{q^2}{6\pi\varepsilon a_0 c^2} \right] v^2 \qquad (8\text{-}13)$$

The total energy of this low velocity electron is the sum of T and the U_0 in Eq. (8-10), i.e.

$$U = \frac{q^2}{6\pi\varepsilon a_0} + \frac{1}{2} \left[\frac{q^2}{6\pi\varepsilon a_0 c^2} \right] v^2 \qquad (8\text{-}14)$$

The second term is the *magnetic energy*. Since it is the energy of motion it is also the *kinetic energy*. The quantity

$$m_0 = \frac{q^2}{6\pi\varepsilon a_0 c^2} \qquad (8\text{-}15)$$

is the *rest mass*.

Multiplying the rest mass Eq. (8-15) by c^2 yields the first term in Eq. (8-14) which is the rest energy. Hence the *rest energy*

$$U_0 = m_0 c^2 \qquad (8\text{-}16)$$

is the so-called *equivalence of mass and energy* for this model of the electron at low velocities. One may then write total energy

$$U = m_0 c^2 + \frac{1}{2} m_0 v^2 \qquad (8\text{-}17)$$

which is the sum of the rest mass energy and the kinetic energy.

8-5 Oblate spheroid electron model

The electron is assumed to be *nonrigid*. When at rest or at low velocities the electron has the shape of a sphere. At high velocities a combination of forces alter the shape of the electron. It takes on the shape of an *oblate spheroid*, having an elliptical shape when viewed from a direction that is transverse to the direction of motion (Fig. 8-1). It is *shortened in the di-*

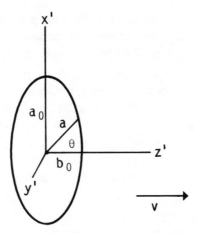

Fig. 8-1 Cross section in the xz plane of an electron moving with v = 0.8c
in the z direction.

rection of motion, making this dimension the semi-major axis
b_0 of the ellipse. The dimension of the major axis a_0 remains
unaltered by motion.

The constant dimension of the major axis results from an
inward *magnetic pinch* on the surface charge that exactly
counterbalances the outward increase in electric tension, as
the transverse E field increases with speed (Fig. 7-1b). The
shortening in the direction of motion results from the reduction, with high velocity, of the electric field in the direction of
motion (Fig. 7-1b) and hence the lessening of outward tension. The binding force pulls that surface of the electron in
until the stable balance is reached at a smaller value of b_0 for
that velocity. For a more complete treatment of this model
one should consult Pemper's Masters Thesis or the related
paper. [3,4]

This elliptical cross section has the shape

$$a = \frac{a_0}{\gamma(1-\beta^2\sin^2\theta)^{1/2}} \qquad (8\text{-}18)$$

As the velocity v of the electron approaches the speed of light c the shape of the electron becomes a thin circular disk of radius a_0 turned sideways to the direction of motion.

The forces acting on the electron are similar to forces that have already been treated. The *magnetic pinch force,* that prevents the electron from increasing the a_0 dimension, is basically the same as the pinch force in Eq. (7-39). Each infinitesimal element of the surface charge on the moving electron has this pinch force acting on it, due to the B field laid down in the S frame by all of the rest of the surface charge on the electron.

The *binding stress* (dF/dA) is similar to that in Eq. (8-8) but incorporates the variable radius vector a of Eq. (8-18) which changes with velocity. The *electric tension* on the surface of the electron is the force per unit area $\epsilon E^2/2$, where E is given by Eq. (7-29) making it a function of velocity.

8-6   Electron energy at high speed

When an electron has been accelerated to a very high velocity the volume of the electron is diminished. Some of its binding field energy has been transferred to the field energy. The electric field energy is greater and is compressed into the transverse area as shown in Fig. 7-1b. Similarly the B field energy is compressed into this transverse region.

The total electron energy U may be written

$$U = U_{bf} + V + T \tag{8-19}$$

as before, but these high velocity expressions are a little more difficult to derive. The results may be summarized as follows:

1) Binding field energy

$$U_{bf} = \frac{q^2}{24\pi\epsilon a_0 \gamma} \tag{8-20}$$

2) Electric field energy

$$V = \frac{\gamma q^2}{6\pi\epsilon a_0} \left[\frac{3}{4} - \frac{\beta^2}{4}\right] \tag{8-21}$$

3) Magnetic field energy

$$T = \frac{\gamma q^2}{6\pi\varepsilon a_0}\left[\frac{\beta^2}{2}\right] \qquad (8\text{-}22)$$

Summing these three types of energy one has for the total energy

$$U = \frac{\gamma q^2}{6\pi\varepsilon a_0}\left[\frac{1}{4\gamma^2} + \frac{3}{4} - \frac{\beta^2}{4} + \frac{\beta^2}{2}\right] \qquad (8\text{-}23)$$

By algebraic manipulation the bracket term reduces to one, simplifying the expression for the *total energy of the electron* to

$$U = \gamma\left[\frac{q^2}{6\pi\varepsilon a_0}\right] \qquad (8\text{-}24)$$

But the bracket term is the *rest mass energy*

$$U_0 = \frac{q^2}{6\pi\varepsilon a_0} \qquad (8\text{-}25)$$

and the rest mass is

$$m_0 = \frac{q^2}{6\pi\varepsilon a_0 c^2} \qquad (8\text{-}26)$$

hence one may also write

$$U = \gamma m_0 c^2 \qquad (8\text{-}27)$$

for the *total energy of the electron*. This is the same equation for total energy as that in special theory of relativity, but it has been derived by this *new theory of electrodynamics* without the Einstein postulates. There is no need for the hypothetical fourth dimension, or contracting meter sticks, or time dilation. These results have been obtained by the common-sense type of clocks and geometry.

CHAPTER 9
Alternative to Einstein's General Theory of Relativity

9-1 Static vs. dynamic gravitational field

Chapter 4 gives a *static* electric theory of gravitation. That theory incorporates electric theory into Newton's universal law of gravitation. It is not an alternative to Einstein's general theory of relativity, his theory of gravitation. Einstein's theory of gravitation takes into account the propagation of gravitational waves. Newton's gravitation law is an action-at-a-distance law. It does not take into account the radiation or absorption of gravitational waves. In that sense our initial electric theory of gravitation is one of electrostatics. We now present a dynamic theory, one that includes a dynamic gravitational field.

This alternative to Einstein's general theory yields all of the applications known for Newton's law of gravitation plus the "expected" dynamical effects of gravitational waves and radiation. These are minute effects that Newton failed to include. Reasoning that gravitational waves are similar to electromagnetic waves, we use in this new theory of gravitation the same analytical form as Maxwell used in his electromagnetic theory of light. This theory was developed in collaboration with Raymond J. Upham, Jr. (1976) and was published in the Creation Research Society Quarterly under the title "A New Theory of Gravitation: An Alternative to Einstein's General Theory of Relativity."[1] Those who may be interested in a more complete mathematical and physical development of the theory should read the technical paper.

9-2   Gravitation vs. electromagnetic analogy

Our dynamic theory of gravitation assumes that the gravitational field has four field vectors d, g, h, and b. By analogy with the electromagnetic theory they correspond to the four electromagnetic field vectors D, E, H, and B. The gravitational field vector g is the conventional g field, the acceleration of gravity.

In electromagnetic theory the electric field vectors D and E are connected by the equation

$$D = \varepsilon E$$

and the magnetic field vectors B and H by the equation

$$B = \mu H.$$

By analogy the following equations are assumed for the gravitational field

$$d = \varepsilon_g g \tag{9-1}$$

$$b = \mu_g h \tag{9-2}$$

where ε_g and μ_g are the proportionality constants, the gravitational *permittivity* and the gravitational permeability. The vectors h and b are the gravitational equivalent of *magnetic field* and *magnetic induction*. It takes a moving mass m to produce h just like it takes a moving charge q to produce a magnetic field H.

9-3   Maxwellian type gravitational equations

The assumption is made that in the gravitational field there are four field equations that are analogous to Maxwell's four field equations. The following sets of equations illustrate that analogy.

Electromagnetic	Gravitational	
$\text{div } D = \rho_e$	$\text{div } d = -\rho$	(9-3)
$\text{div } B = 0$	$\text{div } b = 0$	(9-4)
$\text{curl } H = J + \dot{D}$	$\text{curl } h = J_m - \dot{d}$	(9-5)
$\text{curl } E = -\dot{B}$	$\text{curl } g = \dot{b}$	(9-6)

The negative sign has to be introduced into (9-3) because "like" masses attract, whereas like charges repel. The ρ stands for mass density, mass per unit volume. The J_m is the *mass current density,* the mass per second passing through a unit area. It is analogous to the electric current density J in Maxwell's equation.

9-4 Wave propagation

The dot above a vector means the partial derivative of that vector with respect to time. For example: $\dot{d}$ is the rate of change of field vector d with respect to time. The $\dot{b}$ is the rate of change of the b with respect to time. The $\dot{d}$ is analogous to the electric displacement current density $\dot{D}$. A time rate of change in the field vectors is required to have radiation of gravitational waves. In radiation of gravitational waves, the mass does not propagate through space. It is the fluctuation of d that is propagated through space. Gravitational waves can propagate through empty space where there is no mass.

The wave equation can be derived from the above equations in the same way that one would derive the electromagnetic wave equation from the above electromagnetic equations. The wave equations have the same form and the same type of equation for the speed of the wave. The speed of the gravitation wave in free space is

$$c = \frac{1}{\sqrt{\mu_g \epsilon_g}} \tag{9-7}$$

The gravitational wave equation for g is

$$c^2 \nabla^2 g = \frac{\partial^2 g}{\partial t^2} \tag{9-8}$$

as would be expected. The wave equation for h is the same equation with h replacing g. From those wave equations one can deduce all the wave equation solutions by the same methods as in electromagnetic waves.

9-5 Gravitational Poynting vector

Applying the same mathematical methods to the wave equations as in electromagnetic theory one can deduce a gravitational Poynting vector that shows the direction and intensity (power per unit area) of the gravitational wave.

$$\text{Poynting vector} = g \times h \text{ watts/meter}^2 \tag{9-9}$$

This means, of course, that this gravitational wave is a transverse wave with the g and h vectors at right angles to each other in plane wave propagation.

Knowing the gravitational constant G one may evaluate the gravitational permittivity as follows. Taking the gravostatic case in which Newton's second law holds

$$F = mg \tag{9-10}$$

and Newton's universal law of gravitation holds so the gravitational attraction on mass m due to mass m' is

$$mg = \frac{Gmm'}{r^2} \tag{9-11}$$

and from those

$$g = \frac{Gm'}{r^2} \tag{9-12}$$

From Eq. (9-3) and Eq. (9-1) one can show that the g field at distance r from spherical mass m' is given by the equation

$$g = \frac{m'}{4\pi\varepsilon_g r^2} \qquad (9\text{-}13)$$

From the last two equations one has

$$\varepsilon_g = \frac{1}{4\pi G} \qquad (9\text{-}14)$$

Knowing the value of the gravitational constant one can evaluate the gravitational permittivity.

9-6   Electromagnetics in the gravitational wave

Up to now we have not introduced electrical quantities into our dynamic gravitational theory, only *mass*, and some equations that are *analogous* to electromagnetic theory equations. We have previously assumed, however, that all mass is made up of elementary plus and minus particles, ordinarily electrons and protons in equal amounts. In Sec. 13-4 we shall show that even the neutron is composed of a proton and electron. Hence, every time we deal with mass we are dealing with electromagnetics whether we realize it or not.

We now refer back to Chapter 4, Electric Theory of Gravitation, and incorporate some of those electric concepts into our gravitational waves. Because they contain both plus and minus charges, every ordinary body has associated with it dormant component electric fields of both plus and minus fields. An acceleration of the mass causes electromagnetic radiation of waves in both of these dormant fields. These fields' waves being equal and opposite in value cancel except when they impinge on the surface of an elementary charge. There the asymmetric nonlinearity described in Chapter 4 yields a net attraction on the charges, the conventional gravitational attraction. Under this theory the gravitation wave travels with the same speed as an electromagnetic wave since it is composed of two equal but opposite electromagnetic waves traveling together. These waves would not be detectable under ordinary circumstances because the two component electric waves cancel out except for nonlinear effect at the elementary particles on which these waves impinge. That unbalance yields an extremely small attraction.

This gravitational force is so infinitesimal that it cannot be measured in a situation such as that just described. Scientists have been attempting to measure the wave from accelerating large astronomical objects such as double stars. Even that effect is so small that those experiments are not yet sensitive enough for real confidence in the presumed results. However, it is physically reasonable to expect disturbances in the gravitational field to be propagated as a wave carrying energy like ordinary electromagnetic waves. The reason they are so difficult to detect is that these two component electromagnetic waves cancel except for a rather minute unbalance force exerted on the elementary particles of the mass on which these waves impinge.

9-7   Gravitational analogy to magnetic dipoles

There is a gravitational analogy to the *magnetic moment* of a ring of current I enclosing area A. In the electric case, the magnetic moment is the electric current times the enclosed area. The analogous gravitational moment is for an extremely thin ring of mass rotating about its axis. The mass current is the mass times frequency in this ring. So mass current may be written as

$$I_m = \frac{m\omega}{2\pi}$$

where frequency is expressed in terms of angular velocity. The gravitational moment M_g is the current times the area enclosed by the ring, namely,

$$M_g = \frac{m\omega r^2}{2} \tag{9-15}$$

Here we have an equation for a gravitational dipole "magnet." M_g is its dipole moment.

Two such rotating mass rings may have in addition to their gravitational attraction a slight "magnetic" repulsion or attraction, depending on the orientation of these gravitational magnets. Obviously this treatment can be extended to the gravitational "magnetic" effect of rotating stars.

9-8   Analogous gravitational Lorentz force

In the electromagnetic case the electric and magnetic force on a charge is given by the Lorentz force equation

$$\mathbf{F} = q\mathbf{E} + q\mathbf{v} \times \mathbf{B} \qquad (9\text{-}16)$$

By analogy the gravitational force on a moving mass is given by the equation

$$\mathbf{F} = m\mathbf{g} + m\mathbf{b} \times \mathbf{v} \qquad (9\text{-}17)$$

The reversed order of **b** and **v** in the gravitational case is related to the sign change required in gravitation. "Like masses" attract rather than repel as would be the case in electricity. However, this will give the direction of the gravitational force on moving mass m with the first term being the static gravitational attraction analogous to the electrostatic force on a charge. The second term on the right is the one that is analogous to the magnetic force.

As mentioned in the previous section, this gravitational "magnetic force" applies to the force between two rotating masses. This is analogous to the force between two magnetic dipoles and there can be a "magnetic" repulsion in such a gravitational case. But an actual calculation of the magnitudes will show that the conventional gravitational *attraction* will greatly exceed this repulsion.

One should not get his hopes up for this slight gravitational repulsion force to make it possible to have flying saucers with rotating masses. It just will not work. There might, however, be a hypothetical elementary particle case where a rotating system could have its periphery moving with speeds in excess of the speed of light. Two elementary particles of neutral charge, such as neutrons, might conceivably have a net repulsion. However, that is mere conjecture at this time.

References

1. Barnes, Thomas G., and Raymond J. Upham, Jr. Another theory of gravitation: an alternative to Einstein's general theory of relativity. *Creation Research Society Quarterly,* Vol. 12, March 1976, pp. 194-197.

CHAPTER 10

Hydrogen Atom Components: Electron and Proton

10-1 Diamagnetism

An inherent magnetic property of all bodies is *diamagnetism,* although it may not necessarily be the dominant property. The dominant magnetic properties that cause bodies to be attracted toward a magnet may be paramagnetism or ferromagnetism. Even in those bodies there is a diamagnetic effect, a slight repulsion effect, that is weaker than the attraction effect. *A diamagnetic body when placed in an inhomogenous magnetic field tends to be repelled toward the weaker regions of the magnetic field.*[1] This repulsion force is a natural electromagnetic phenomenon. The external magnetic field induces an electric current in the electrons within the atomic structure. According to *Lenz's law* the induced current sets up a magnetic field that opposes the inducing field. This induced repulsion force remains as long as the body is in the external magnetic field. That current remains because these internal electron currents have no resistance losses.

Some of the diamagnetic materials are: arsenic, bismuth, copper, diamond, gold, lead, mercury, nitrogen gas, silver, sulfur, and rock salt. Although the repulsion force in these materials is relatively weak, it can be very strong in an electron. The reason the repulsion force in the electron can be so strong is two-fold: 1) it is an ideal diamagnetic body; 2) it can be located in regions where the electromagnetic induction is very strong.

10-2   Diamagnetic property of an electron

Contrary to the assumption in quantum theory, we assume that *the electron has no intrinsic spin*. All of the spin of an electron is assumed to be due to *magnetic induction* from some external magnetic field. There is no fixed (quantum) value of spin. The amount of spin depends on the amount of magnetic induction. The spinning electron (rotating charge) is an ideal electric current that has no resistance, no ohmic loss of energy. The induced electric current *always* has such a direction that the magnetic field which induced it exerts a *repulsion force* on the electron.

This magnetic repulsion effect is the *diamagnetic* property of an electron. This repulsion force prevents the electron from falling into the proton. If it were not for the diamagnetic property of the electron it would smash into the proton and be "discharged" by the positive charge of the proton. It is the balance between this magnetic repulsion force and the electric attraction force on the electron that holds together the electron and proton in the hydrogen atom.

The proton is assumed to have an intrinsic spin and associated magnetic field. That magnetic field induces a spin into the electron and sets up the magnetic repulsion between the proton and the electron. The electric attraction between the positive proton and negative electron pulls the electron into the hydrogen atom. The magnetic repulsion force stops it at the right distance to form the atom without the negative electron ever touching the positive proton. Were it not for this diamagnetic property of the electron, there would be no hydrogen atom, nor any other atom.

This important diamagnetic property of the electron itself
has been ignored too long. The reason it has been ignored is
the popularity of quantum theory. Quantum theory does not
allow an electron to have an induction spin current. Quantum
theory is not concerned with physical models that obey the
laws of electromagnetic theory. Nevertheless, an actual pure
electric charge that is free to spin when it moves into a magnetic field will spin and have this diamagnetic property. That
is an inescapable deduction from the laws of electromagnetic
theory. There should be no reason for abandoning those laws
when constructing a model of the electron. Without a physical
model one is hard pressed to provide a logical reason for the
physical processes involved.

It was logic based on the laws of electricity and magnetism
applied to a model of the electron, a model somewhat similar
to the model we propose, that led H.A. Lorentz to skepticism
of the quantum spin (see Sec. 6-4). He considered it to raise
a problem with the presumed equivalence of mass and energy,
a point which we have already mentioned. He did not introduce the concept of the diamagnetic property of the electron
but logical reasoning of the classical type employed by Lorentz
does favor a model of the electron that has this diamagnetic
property. Logical reason associated with physical phenomena
does not favor the quantum spin. Quantum theory is so nebu-
lous in regard to physical analysis that it does not really consider the spin to be an actual rotating electric charge. It completely abandons the beautiful *logic* of classical physics, that
fruitful domain of science upon which most of the advances
in modern technology are based.

10-3 Proton's intrinsic spin

Since *all* of the spin of an electron is due to the magnetic
induction from an external magnetic field, it is an *ideal diamagnetic body*. The electron would have no spin if it were
taken completely out of an external magnetic field. That is not

a condition that can be achieved experimentally because there will always be *some* magnetic field where an experiment might be performed on an electron. However, that does not alter the conclusion that an electron has that *ideal diamagnetic property,* a property that is essential for our theory of the hydrogen atom.

The proton is assumed to have an intrinsic spin. It always has a magnetic field to induce a spin into the electron when the electric force pulls the electron in toward it. In forming the hydrogen atom by recombination of the negative electron and the positive proton, the proton's magnetic field induces sufficient spin in the electron to yield the proper separation of electron and proton in that atom.

Even though the proton has the intrinsic spin there can also be some magnetically induced spin. The total spin is not a constant. However, the spin rate change due to induction may be small compared with the proton's intrinsic spin.

There is experimental evidence that the proton's spin can be greatly increased under certain extreme conditions. That evidence is found in experiments with proton beams in which the protons have been accelerated to extremely high energies and interact with other particles. In such interactions there is presumably a means of distinguishing between the spin interaction and the translational interaction. If the proton's spin were constant, the *ratio* of spin effect to translational effect would diminish when translational energy is increased.

In his 1979 Scientific American article on The Spin of the Proton, Alan D. Kirsch gives this report on the research at the Argonne National Research Laboratory:

It has long been thought, however, that the influence of spin should decline as the energy of the collision increases. The reasoning behind this assumption is simple: the energy associated with a proton's spin is constant, and so it becomes an ever smaller fraction of the total energy as the collision becomes more violent. . . Only in the past few

years have experimental techniques been devised for test-
ing this assumption. It has turned out to be quite wrong.
The influence of spin does not diminish as the energy of a
collision increases: on the contrary, spin seems to become
more important as the collision becomes more violent.[2]

The *relative* spin effect did not diminish as expected. The spin
must have *increased* in the experiment, contrary to the quantum theory prediction. Presumably the proton did respond to
magnetic induction and the spin increased. That is strong evidence against the quantum theory of spin.

10-4   Potential energy for recombination

Several energy conversions take place when a free electron
falls in toward a proton. Let us first find the electrostatic
potential energy that will be given up during this process.
The potential energy depends only on q^2 (the product of the
charge on the electron and the charge on the proton) and the
distance r between their centers.

The potential energy of the electron is equal to the work
done in separating the electron from the proton, work done
against that electric attraction force. It is convenient to arbitrarily call the potential energy zero when the electron is completely free of the proton's electric field, that is when $r = \infty$.
The loss of potential energy is then given by the equation

$$\text{P.E.} = \frac{-q^2}{4\pi\varepsilon r} \tag{10-1}$$

The minus sign indicates that the potential energy is lost
(given up) by the electron in moving down to within distance r
of the proton.

That potential energy is the source of energy that produces
the recombination of an ionized hydrogen atom, the bringing
back together of the electron and the proton into the atom.

During this process of the electron falling back into the
atom, the potential energy given up will be transformed into
several different forms of energy; kinetic energy (energy of

motion), magnetic energy (spin energy), and some radiation energy (Fig. 10-1). The radiation takes place during the oscillations that occur before the electron settles down. It settles down into a spinning, but nonradiating state (see Sec. 11-2). The result is a hydrogen atom in the ground state.

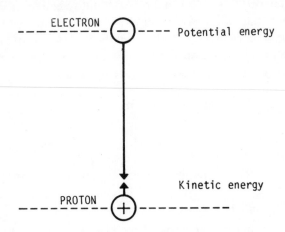

Fig. 10-1 Equal Coulomb attraction force on the electron and proton produces 1,836 times more motion in the electron than in the proton because the mass of the proton is 1,836 times as great as the mass of the electron.

10-5   Deformable electron

It was shown in Chapter 8 that at high speed the feedback shifts the electric field of an electron such that it is weaker in the fore and aft ends and stronger on the transverse sides. This change in the stress forces on the electron causes it to change its shape from a sphere to an oblate spheroid. This means that the electron is a *deformable body* in which its shape can be altered by forces acting on it. It is only when the electron is at rest that it will have the spherical shape. The oblate spheroid shape exists during high speed linear motion.

When the electron falls in toward the strong magnetic field of the proton, a high rate of spin is induced into the electron. The question is: What kind of deformation takes place when the electron gets in very close to the proton? We shall see in Sec. 10-11 that the magnetic field of the proton exerts enough pressure on the electron to prevent the electron from touching the proton. That repulsion pressure puts a deep inward bulge into the electron. The centrifugal force due to the electron's spin also has a part in shaping the electron (see Sec. 11-4).

The possibilities of developing classical models for the neutron and for the hydrogen atom are greatly enhanced by a knowledge of the kind of deformation the electron can have in the hydrogen atom and in the neutron. In Chapter 13 it will be seen that this provides a long awaited answer to how there can be an electron and a proton in a neutron without the electron discharging into the proton. It also provides for a possible hydrogen atom structure that can yield many more modes of vibration, hopefully modes that can radiate the known spectral frequencies of the excited hydrogen atom.

The magnetic induction, spin, and the deformed shape of the electron are not factors in the motion of the electron, as it falls in toward the proton, until it gets relatively close to the proton. Let us first consider the *translational* motions of both the electron and proton. The Coulomb electrostatic attraction that causes the electron to fall in toward the proton exerts an equal and opposite attraction force on the proton. The question that needs to be answered is: How much translational motion does it impart to the proton? We shall consider that in the next section.

10-6   Translational energy delivered to electron and proton
Note from Fig. 10-1 that when an electron falls in toward a proton the potential energy is transformed to translational kinetic energy. Some of that translational kinetic energy is delivered to the electron and some to the proton. By aid of

Newton's laws one can show that the electron receives most of that energy.

From Newton's third law when the electrostatic force F_e pulls the electron in toward the proton, an equal and opposite force F_p pulls the proton in toward the electron. Newton's second law states that the net force on any body equals its mass m times its acceleration a. From those two laws one may write the equation

$$m_e a_e = m_p a_p \qquad (10\text{-}2)$$

where the subscript e associates that quantity with the electron and p with the proton. Solving for the acceleration of the electron

$$a_e = \frac{m_p}{m_e} a_p \qquad (10\text{-}3)$$

This shows that the electron is accelerated $\frac{m_p}{m_e}$ times as much as the proton. The ratio of the mass of the proton to the electron is 1,836. So the electron is accelerated 1,836 times as much as the proton. It will move 1,836 times as far as the proton.

Remembering that the translational energy delivered to a body is the integral of force times distance and that the forces are equal, it follows that 1,836 times as much translational energy is delivered to the electron as to the proton. For most practical purposes one may say that the electron's potential energy is transformed to kinetic energy in this region where the magnetic influence is small.

10-7   Induction of magnetic energy into the electron

As the electron falls in toward the proton the magnetic field of the proton induces a spin in the electron. The diamagnetic property of the electron positions the electron so that it falls in along the plane that bisects the proton and its spin axis is perpendicular to that plane (Fig. 10-2).

The diamagnetic property of the electron orients the spin so

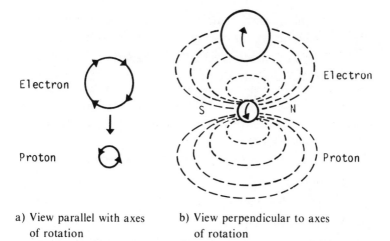

a) View parallel with axes   b) View perpendicular to axes
   of rotation   of rotation

Fig. 10-2 a) Shows the direction of the spin induced in the electron as it
falls in toward the spinning proton. b) Shows the proton's
magnetic flux linking the electron.

that its induced magnet opposes the inward fall toward the
proton magnet. The magnetic dipole of the electron is parallel
to the magnetic dipole of the proton. The north magnetic pole
of the electron is above the north magnetic pole of the proton.
The south magnetic pole of the electron is above the south
magnetic pole of the proton. The repulsion force is in accordance with the simple rule that *like poles repel.*

*The magnetic energy in the electron is equivalent to its ro-
tational kinetic energy* (see Chapter 5). One has a choice of
expressing this energy either in magnetic terms or in the terms
of mechanics. The energy transformation may be written in
terms of the initial and final energy. The translation is an
intermediate process that need not be included.

Potential energy = magnetic energy + radiation energy.

The potential energy lost by the electron

$$\text{P.E.} = \frac{q^2}{4\pi\varepsilon d} \qquad (10\text{-}4)$$

where d is the distance from the center of the proton to the *center of electric potential energy* of the electron.

The radiation energy is equal to the energy it would take to ionize the hydrogen atom:

$$\text{Radiation energy} = qV \tag{10-5}$$

where V is the *ionization potential* of the hydrogen atom and q the electron charge.

The magnetic energy may be written as rotational kinetic energy, namely

$$\text{Rotational kinetic energy} = \frac{I\omega^2}{2} \tag{10-6}$$

where I is the moment of inertia of the electron and ω is its angular velocity. The energy transformation equation may then be written as

$$\frac{q^2}{4\pi\varepsilon d} = \frac{I\omega^2}{2} + qV \tag{10-7}$$

In. Sec. 10-8 we will apply this equation to the falling electron.

In Sec. 11-5 it will be shown that the magnetic energy is contained in the proton's added spin as well as in the electron's spin, giving a more complete solution than that in Sec. 10-8.

10-8   Final position of the electron

We shall assume that the electron's magnetic moment in its equilibrium position in the hydrogen atom has the same value as that found in the technical literature. Equation (10-7) can be used to obtain an estimate of the final position of the electron as it falls into the hydrogen atom. The proton and electron are assumed to be spherical. The distance d is the distance between their centers.

The rotational kinetic energy, Eq. (10-6), will be replaced by its equivalent, the magnetic energy, Eq. (5-21).

$$w_m = \frac{\mu M^2}{4\pi r^3} \tag{10-8}$$

Denoting the radiation energy as w_r Eq. (10-7) may be written
as

$$\frac{q^2}{4\pi\varepsilon d} = w_m + w_r \qquad (10-9)$$

Using the following values: $M = 9.1285 \times 10^{-24}$, $\mu = 4\pi \times 10^{-7}$,
$r = 1.8786 \times 10^{-15}$, yields $w_m = 1.30 \times 10^{-9}$ joules for the *magnetic energy* of the electron. Using $V = 13.599$ volt for the
ionization potential of the hydrogen atom and $q = 1.6022 \times 10^{-19}$ coulomb for the charge on the electron yields, $w_r = 2.18 \times 10^{-18}$ joules. That is the radiation energy lost by the electron
before it settles into its equilibrium state after having fallen
in from free space. Note that this radiation energy is negligible
compared to the magnetic energy induced into the electron.

Equation (10-9) reduces to

$$\frac{q^2}{4\pi\varepsilon d} = w_m$$

Using $\epsilon = 8.85 \times 10^{-12}$ in the evaluation, yields the separation
distance $d = 1.8 \times 10^{-19}$ meter. That is much less than the
radius of the spherical electron.

What that means is that by the time the electron's fall has
been stopped it has been deformed by a very deep dent. It
sinks down to where the proton's magnetic pressure produces
the dent, almost into the center of the electron. The magnetic
field just barely keeps the electron from touching the proton.
This is an approximate solution because the electron is no
longer spherical in this equilibrium position. However, it
gives a classical explanation of how the electron and proton
can be positioned together in the hydrogen atom. A more
complete treatment will be given in Chapter 11.

It can be shown that the spherical electron's magnetic
moment Eq. (5-7)

$$M_e = \frac{q\omega r_e^2}{3} \qquad (10-10)$$

So the angular velocity can be computed from the equation

$$\omega = \frac{3M_e}{qr_e^2} \tag{10-11}$$

Two important physical analyses were not included: 1) The availability of sufficient magnetic flux from the proton to produce that transformation from translational motion of the electron to rotational kinetic energy (magnetic energy). 2) A quantitative indication that there is enough magnetic pressure to keep the proton and electron from touching. Those analyses will be developed in the next sections.

10-9   Required flux linkage in the hydrogen atom

In order for the electron to gain the required magnetic energy, there must be a certain amount of magnetic flux from the proton linking through the electron. This accumulated flux linkage is a result of the induction process that transforms the translational energy into magnetic energy as the electron moves in toward the proton's magnetic field.

We will first find the required amount of magnetic flux linkage ϕ_ℓ, the flux through the electron. Then we will find the total amount of magnetic flux ϕ_t from the proton. By comparison of these two values we can determine whether or not the proton has enough total flux ϕ_t to provide the required flux linkage ϕ_ℓ.

The total flux must of course be greater than the amount that links through the electron. Then by knowing what fraction of the total must link through the electron we get some insight into the required position and shape of the electron in the hydrogen atom.

We have already seen in Eq. (10-8) that the required magnetic energy

$$w_m = \frac{\mu M_e^2}{4\pi r^3} \tag{10-12}$$

where the subscript e associates that quantity with the elec-

tron. It can be shown that the energy induced by the flux linkage is

$$\text{Flux linkage energy} = \frac{\phi_\ell \, I_e}{2} \tag{10-13}$$

where I_e is the electron spin current and can be expressed as
$q\omega / 2\pi$.

Equating the energies in Eqs. (10-12), and (10-13) and making
the substitution from Eq. (10-11) yields the equation for the
required flux linkage

$$\phi_\ell = \frac{\mu M_e}{3r_e} \tag{10-14}$$

which can be evaluated from the known values: $M_e = 9.285 \times 10^{-24}$, $r_e = 1.879 \times 10^{-15}$, and $\mu = 4\pi \times 10^{-7}$. The required value
of flux linkage

$$\phi_\ell = 2.08 \times 10^{-15} \text{ weber.}$$

The total magnetic flux from the proton is found from the
B field inside of the proton and the area. The internal B field
in the spinning proton is constant and given by the equation

$$B = \frac{\mu M_p}{2\pi r_p^{\,3}} \tag{10-15}$$

The total flux is equal to the product BA where A is the cross
section area of the proton. So

$$\phi_t = \frac{\mu M_p}{2r_p} \tag{10-16}$$

Using the values of magnetic moment and radius for the pro-
ton, $M_p = 1.4106 \times 10^{-26}$ ampere meter^2, and $r_p = 1.023 \times 10^{-18}$
meter in Eq. (10-16) the total magnetic flux of the proton is

$$\phi_t = 8.66 \times 10^{-15} \text{ weber.}$$

Note that the total magnetic flux of the proton is only slightly more than four times the required flux linking the electron. That tells us that the electron must be shaped and positioned so as to "intercept" nearly one fourth of the total flux.

10-10 Special features of the proton

Our proton model is considerably smaller, radius of approximately 10^{-18} meter, than the size usually given in the literature, radius of approximately 10^{-15} meter. This smaller dimension made it possible to use the same electric equation for the rest mass of the proton as that for the rest mass of the electron, namely

$$m = \frac{\mu q^2}{6\pi r} \qquad (10\text{-}17)$$

The proton being smaller than the electron makes sense because a small radius with the same charge yields a stronger electric field in the region near its surface and hence more inertial mass.

Similarly, the smaller size proton with the same magnetic moment, yields a stronger magnetic field near its surface. Its magnetic dipole field varies inversely as the *cube* of r, as compared with its electric field varying inversely as the *square* of r. The magnetic effect, the repulsion, dominates in the region near the proton where r is small. That is important in keeping the electron from touching the proton in this model of the hydrogen atom where that spacing is very small. It is this repulsion force that causes the bulge in the electron that prevents it from touching the proton.

Another advantage of the small proton is that it yields a larger total magnetic flux as can be seen from Eq. (10-16). As seen in the previous section, this small size proton produces sufficient magnetic flux to induce the required spin and associated magnetic energy in the electron to provide the required energy transformation. That makes possible a classical model of the hydrogen atom. A larger proton would not have supplied enough repulsion to keep the electron from falling on into contact with the proton.

10-11   Speeds exceeding speed of light

The feedback theory developed in Chapters 7 and 8 can be applied to certain types of motion in which there can be speeds in excess of the speed of light. The outer rim of our model of the proton moves with speed that is much greater than the speed of light. For example, the equation relating the angular velocity and the magnetic moment of the proton is

$$\omega = \frac{3\,M}{qr^2} \tag{10-18}$$

The rim velocity is simply the angular velocity times the radius

$$v = \frac{3\,M}{qr} \tag{10-19}$$

Using the previously mentioned values of the proton's charge, magnetic moment, and radius yields the speed

$$v = 2.58 \times 10^{11}\ \text{meter/sec}$$

which is 861 times the speed of light.

According to the feedback theory, as seen in Chapter 8, the proton or electron can not move with translational speed in excess of the speed of light because of the feedback force from the changing magnetic field at fixed points in the medium. In the case of rotational motion of a body such as the proton there is no disturbance in the B field as long as the rotation rate is constant. Hence, there is no feedback force or torque to prevent the proton's edge from moving with speed greater than the speed of light. It is only the feedback force and the associated build up of field energy such as that discussed in Chapters 7 and 8 that limits the speed to less than the speed of light. Translational motion of a body is limited to less than the speed of light. Rotational motion of a continuous body, such as a proton, is not limited in any of its parts to speeds less than the speed of light.

In fact, it is this speed in excess of the speed of light at the rim of the proton, that enables it to have a greater magnetic repulsion force on the nearby portion of the electron than the electric attraction force. (See Chapter 13).

References

1. American Institute of Physics handbook, third edition, 1972, p. 5-3.
2. Kirsch, Alan D. The spin of the proton. *Scientific American,* May 1979, p. 68.
3. Barnes, Thomas G. New proton and the neutron models, *Creation Research Society Quarterly,* Vol. 17(1), pp. 42-47, June 1980.

CHAPTER 11
A Classical Hydrogen Atom Configuration

11-1 Classical defects in the Bohr model

The Bohr model of the hydrogen atom is illustrated in Fig. 11-1. In the stable state of the hydrogen atom the electron moves around the proton in a circular orbital motion. According to Bohr in the ordinary state of the atom that orbit must have a fixed value of radius, which turns out to be about 5.29×10^{-11} meter.

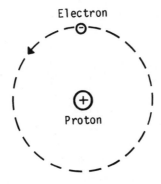

Fig. 11-1 Bohr model of the hydrogen atom.

There are four classical defects of that model of the hydrogen atom:

Defect number 1.

There can be no *stable* state of an electron moving freely in a circular orbit. That electron would be continuously radiating electromagnetic energy. That continuous loss of energy would cause it to spiral into the proton and the hydrogen atom would die. The reason the electron would lose energy is this: In that orbital motion there is centripetal acceleration (acceleration toward the center). Acceleration of an electron *always* causes it to radiate electromagnetic energy. That is the way a radio wave is produced. The electrons in the antenna are accelerated.

Defect number 2.

As an electron falls in toward the proton to form the hydrogen atom, there is no classical reason why it would always "choose" that particular orbit. For example, it could fall in closer and orbit in a circle of smaller radius, or in any number of elliptical orbits.

Defect number 3.

Bohr did not include a spin of the proton or electron in his model. His model has no magnetic means of preventing the electron from falling directly into the proton. Why wouldn't some of the electrons fall directly into the proton in the hydrogen atom?

Defect number 4.

In the Bohr theory, when the hydrogen atom is excited the electron is temporarily raised to an orbit of larger radius. Only certain larger radii are "allowed." The allowed radii are governed by a quantum rule. When the electron falls from an orbit of larger radius to one of smaller radius, a quantum of energy is radiated. That quantum of energy w is specified by the quantum equation

$$w = h\nu \qquad (11\text{-}1)$$

which turns out to be the difference in the kinetic energy of the electron in these two different orbital motions. The defect is that this Bohr model provides no *classical* mechanism for generating an electromagnetic wave of that frequency. It is strictly a *quantum postulate*.

There is no common sense reason for abandoning the basic classical physics principles which expose these defects in the Bohr theory. His theory contains a mixture of classical theory and quantum theory. Where classical theory is violated quantum theory was developed to "remedy" the problem.

11-2   A classical remedy for three of Bohr's problems

The reason the orbiting electron will radiate is that it sets up *disturbances* in its electric and magnetic field. Those disturbances move out as radiation energy. If one pictures the electric field of the electron by a set of radial lines extending out from the electron, when the electron is accelerated it will set up kinks in those lines and the kinks will move out with the speed of light. (Fig. 11-2). This is an illustration of the pro-

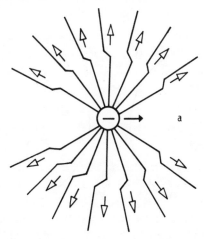

Fig. 11-2 Radiation is produced by acceleration of the electron. The acceleration sets up disturbances in the electric field that move out with the speed of light. The disturbance is shown as kinks in the electric field lines.

cess of electromagnetic radiation. The electron is continuously accelerated toward the center when it is in circular orbital motion. Radiation would be continuously taking place because acceleration continuously sets up the disturbances in the field lines as the electron moves in its orbital path.

One rule as to whether or not there will be radiation is this: *If there is a net disturbance set up in the electric or magnetic field, there will be radiation. If there is no net disturbance set up in the field, there will be no radiation.*

An electron or proton spinning at a constant angular velocity meets that condition of no net disturbance in its electric or magnetic field. Hence there is no radiation. A model of the hydrogen atom that incorporates a spinning electron and spinning proton in its stable state does not have that defect of the Bohr atom.

It is not necessary for the rotating electron to be spherical to meet the non-radiating condition. In the proposed model

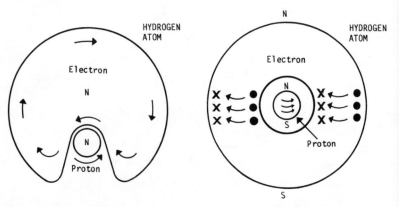

a) Section through the depression   b) Looking into the depression

Fig. 11-3 Hydrogen atom consisting of a spinning proton and revolving electron in the configuration shown. The N and S indicate magnetic north and south polarities.

there is an indentation in the electron due to the close spacing of the proton. However, the same configuration of the electron exists so long as it is in its ground state, even though there is the rotational motion in the electron. There is no disturbance in the field and therefore no radiation.

In the next section we will show that the diamagnetic property of the electron will cause it to fall into the position and orientation shown in Fig. 11-3. This process eliminates the need for electron orbits. The spins and this orientation supplant those orbits. It will also be shown that their magnetic force due to these spins provide the needed separation of the positive and negative charges of the proton and electron. Hence, the Bohr defects number 2 and 3 are also eliminated in this proposed classical model of the hydrogen atom.

11-3 Diamagnetic repulsion guaranteed

One might ask why the electron's magnet can always be oriented such that the proton's magnet tends to repel it as it falls into the hydrogen atom. For example, if the electron's magnet is parallel to the proton's magnet in the repulsion orientation (north pole over north pole and south pole over south pole) why doesn't the electron magnet rotate 180 degrees about its magnetic axis such that the two magnets attract rather than repel?

That reorientation and attraction would happen if the electron and proton were *permanent* magnets. There would be instability in repulsion orientation of two permanent magnets. They would tend to reorient themselves into the attraction orientation (south pole over north pole, etc.).

Fortunately that will not happen in the hydrogen atom because of the ideal diamagnetic property of the electron. The proton's magnetic field will always induce a current (spin current) in the electron that produces an electron magnet that opposes the proton's magnetic flux that links the electron. *The electron's spin current is always a function of the proton's magnetic flux through the electron.* That guarantees the dia-magnetic repulsion.

If any other magnetic orientation is taken, it yields more magnetic flux through the electron and hence a force or torque opposing that orientation. If the electron magnet rotates slightly out of its parallel orientation with the proton magnet, more of the proton's magnetic flux links it. That induces the corrective effect that opposes that "misorientation." That diamagnetic property is not only one that helps hold the electron in the atom, it also provides a *restoring torque* that can, under certain conditions, yield sinusoidal oscillations when there is an "excitation" of this type.

One fact should be reemphasized. The electron's spin current is always a function of the proton's magnetic flux through the electron. There is no spin current other than that which has been induced by the flux linkage that exists at that time. Mathematically this is equivalent to a definite integral. It depends only on the end points: 1) The lower limit is zero linking flux and zero spin. 2) The upper limit is the present linking flux and associated spin current.

This means that the present flux linking the electron will cause it to "seek" its optimum diamagnetic position and orientation. That is the diamagnetic tendency to be repelled toward the weaker regions of the field. That is the mechanism that puts the electron in its proper position in the hydrogen atom.

11-4   Stable state of the hydrogen atom

In Sec. 10-9 it was shown that about one fourth of the magnetic flux from the spinning proton must link through the electron to provide sufficient magnetic induction energy in the electron for it to have the rated value of magnetic moment, a value that has presumably been determined experimentally. It was shown in Sec. 10-8 that the electron had to be very close to the proton to meet the energy conditions imposed by the law of conservation of energy. The required electric separation of the electron from the proton is taken care of by the strong magnetic field of the proton in that critical region near the proton.

It appears then that the electron needs to have in it an indentation containing the proton and have a shape something like that in Fig. 11-3. The size, shape and position of the electron shown in this figure seems to meet the condition of intercepting one fourth of the magnetic flux from the proton. The magnetic flux density B of the proton's magnetic dipole diminishes inversely with the distance from the proton. The electron is large enough to intercept the required amount of flux, flux in the near field. Most of the flux that does not link through the electron is that between the electron and proton. It is that flux that forms the needed magnetic barrier between the proton and electron to hold them apart.

In Sec. 13-3 it will be shown that the magnetic force between the surfaces of the proton and electron is so strong that their positive and negative charges will be kept apart. That strong force results from the fact that their rim speeds exceed the speed of light.

11-5   Conservation of angular momentum in the hydrogen atom

According to the law of conservation of angular momentum, the total angular momentum remains constant in a system that has no external torque acting on it. Angular momentum is a vector quantity. The sum remains constant when a negative vector increases the same amount as a positive vector increases. In the hydrogen atom the electron and proton spin in opposite directions as illustrated in Fig. 11-4. Their angular momentum vectors point in opposite directions. One is positive; the other is negative. When the *absolute* magnitude of one vector increases, the absolute magnitude of the other increases an equal amount.

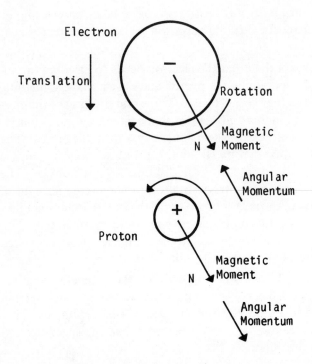

Fig. 11-4 Relative directions of rotation, magnetic moment, and angular momentum of the electron and proton as the electron falls in toward the proton.

Hence, as the electron falls in toward the proton the magnitude of the angular momentum of both the electron and proton increase. Not only does the electron take on more angular momentum, the proton takes on an equal amount of angular momentum.

We assume that the proton had an intrinsic angular momentum $I_p\omega_0$ to start with. It then takes on the above mentioned additional angular momentum. The law of conservation of angular momentum may be written as

$$I_e\omega_e = I_p(\omega_p - \omega_0) \tag{11-2}$$

where the moments of inertia I_e and I_p of the electron and proton are assumed to be constant.

The ratio of the moments of inertia of the spherical electron and proton can be shown to be (See Eq. 5-17)

$$\frac{I_e}{I_p} = \frac{r_e}{r_p} \tag{11-3}$$

because the moment of inertia is proportional to the product of mass times the square of the radius and the mass is inversely proportional to the radius.

In view of Eqs. (11-2) and (11-3) the change in angular velocity for the proton

$$\Delta\omega_p = \frac{r_e}{r_p} \omega_e \tag{11-4}$$

Recalling from Eq. (10-11) that

$$\omega = \frac{3 M}{qr^2} \tag{11-5}$$

and applying that to the electron for ω_e in Eq. (11-4) and to the final ω_p of the proton, the ratio of increase in the angular velocity of the proton

$$\frac{\Delta\omega}{\omega_p} = \frac{r_p M_e}{r_e M_p} \tag{11-6}$$

From the values $r_p = 1.023 \times 10^{-18}$, $r_e = 1.87 \times 10^{-15}$, $M_e = 9.285 \times 10^{-24}$, and $M_p = 1.41 \times 10^{-26}$

$$\frac{\Delta\omega}{\omega_p} = 0.360$$

meaning that the angular velocity of the proton increased by roughly 1/3 in the hydrogen atom from its intrinsic angular velocity. From Eq. (11-5) the angular velocity of the electron and the proton in the hydrogen atom would be

$$\omega_e = 4.97 \times 10^{25} \text{ and } \omega_p = 2.52 \times 10^{29}$$

if they were spherical and had the magnetic moments listed above.

The *intrinsic* spin angular velocity ω_0 of the proton equals the ω_p in the hydrogen atom minus the $\triangle\omega_p$. Evaluating from the above data, the *intrinsic spin angular velocity of the proton*

$$\omega_0 = 1.61 \times 10^{29} \text{ rad/sec}.$$

Similarly, the intrinsic magnetic moment of the proton is the fraction 0.64 of the conventional value which we have used in the hydrogen atom. Thus the intrinsic magnetic moment of the proton is

$$M_0 = 9.024 \times 10^{-27} \text{ amp m}^2$$

The derivation in this section does not take into account the distortion of the electron into a nonspherical body and the complex internal motion in the electron. Those phenomena would certainly alter the angular momenta of the system. The derivation must be considered to be only a qualitative illustration of some of the physical phenomena involved.

The present status of the hydrogen atom model is tentative and incomplete. There are too many unknown factors involved to make the model quantitative at this stage of development. Nevertheless, this approach to the problem opens new possibilities for a classical model of the hydrogen atom that had not been taken into account in the historical efforts to develop a model.

One logical test of the plausibility of this model of the hydrogen atom is to see whether or not two of these atoms might consistently be put together into a hydrogen molecule. That will be considered in the next section.

11-6   Classical model of the hydrogen molecule

The simplest molecule is the hydrogen molecule. It consists of two atoms of hydrogen. The two protons are located in the nucleus. The two electrons are located outside of that nucleus.

One requirement of any model of the hydrogen molecule is to have some force to hold the protons in the nucleus. Because they have like charges (both positive) they tend to repel each other. It is in fact a very strong repulsion force that obeys Coulomb's law.

Another requirement of any model of the hydrogen molecule is to have some force to prevent the electrons from being pulled into contact with the protons. The electron and proton have unlike charges and unlike charges attract. Similarly, this Coulomb attraction force is a strong force within the molecule.

The required force to hold the protons within the nucleus in the model is a magnetic attraction force. The required force to prevent the electrons from being pulled into the protons is a magnetic repulsion force.

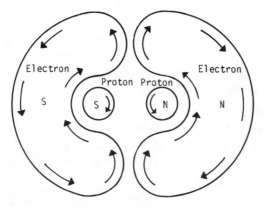

Fig. 11-5 Hydrogen molecule contains two hydrogen atoms so oriented that the magnetic forces hold the protons within the nucleus and prevent the electrons and protons from coming together. (middle cross section)

Figure 11-5 illustrates our proposed model of the hydrogen molecule. Two hydrogen atoms are located close together and oriented as shown. The proton of one atom is near the proton of the other atom. The spins of these two protons are in opposite directions. That makes them have opposite magnetic poles on the top side and opposite magnetic poles on the bottom side. The result is a magnetic attraction force strong enough to overcome their electrostatic repulsion.

As seen in the figure, the electron in one atom has its north magnetic pole opposite the north magnetic pole of its companion proton, yielding a magnetic force that keeps the electron from falling into the proton. Similarly, the other electron in the figure has its south magnetic pole opposite its companion proton's south magnetic pole. This provides the required repulsion force to hold those two opposite charges apart.

Actually the more important point is the distribution of magnetic flux. The small space between the proton and its corresponding electron has a concentration of magnetic flux. This magnetic field acts on the surface currents in these adjacent spinning charges so as to repel enough to equalize the electric attraction between these two unlike charges. The magnetic field pattern between the two protons is such as to cause an attraction on the surface currents of those spinning charges, balancing out their electrostatic repulsion.

The current flow in that portion of the two opposite electrons that might ordinarily be repelled are held in position by a magnetic pinch effect since both surface currents are flowing in the same direction on that portion of the electrons.

This qualitative test of the two classical hydrogen atoms forming a hydrogen molecule is very encouraging. There is, of course, a need to make a quantitative analysis of this model. That would answer questions such as the precise configurations, spacings, and "binding" force and energy.

A promising alternative to the spherical electron in Fig. 11-3 is an *electron ring* of dimensions comparable with the orbit of the Bohr hydrogen atom. In Fig. 11-5 the hydrogen molecule has two such electron rings.

CHAPTER 12
Clock Error —
Not Time Dilation

12-1 Around the world clock flights

A famous experiment by Hafele and Keating[1] has frequently been cited as proof of time dilation. According to Einstein's special theory of relativity, time runs slower in a moving frame of reference. Clocks flown around the world eastward, according to that reasoning, would be moving faster than clocks flown around the world westward, because the earth is rotating eastward. The clock's total speed is the earth's eastward speed plus the aircraft speed on the eastward flight and minus the aircraft speed on the westward flight.

The clocks used were atomic clocks in hopes of getting enough precision to measure the predicted change. The eastward flying clocks were expected to run slower than standard time. The westward flying clocks were expected to run faster than standard time. Hafele and Keating claimed to have succeeded in measuring the time dilation predicted from Einstein's relative time equation. This experiment is widely heralded as additional confirmation of Einstein's special theory of relativity.

12-2   Essen's challenge

Dr. L. Essen has challenged the claim of Hafele and Keating that their results confirmed the Einstein time dilation.[2] No one is better qualified to make this challenge. Dr. Essen is an international authority on the velocity of light measurement, the inventor of the Caesium clock (the atomic clock used in the experiment), and a Fellow of the Royal Society.

It was through a letter from Professor G. Burniston Brown,[3] Department of Physics and Astronomy, University College London, that the author learned that Essen had come to a different conclusion on this experiment. Here is what Brown stated about Essen in that letter:

> He made a re-calculation of the round-the-world clock experiment, using *all* the data published which the original investigators did not use (why not, we don't know). He found no evidence for time dilation,. . .

Essen sent a letter to *Nature* reporting his analysis of the data. Nature refused to publish his letter. Fortunately, that letter has finally been published in the Creation Research Society Quarterly.[4] In that letter Essen analyzes the data in light of his extensive experience with the atomic clocks, giving a detailed explanation of the short comings of the Hafele and Keating paper. He points out:

> The authors then proceed to make a statistical analysis of the frequency comparisons made between the clocks to obtain their final results. No details of these comparisons are given, but the analysis is based on the assumption that the frequency variations are random in nature, which appears to be unlikely and not in accord with my own experience.

Essen states further:

> In their theoretical discussion the authors ignore detailed and fully documented criticisms of Einstein's relativity theory which have been made and have not been refuted.[5, 6]

I suggest therefore that the theoretical basis of their predic-

tions needs careful scrutiny and that the experimental results given in their paper do not support these predictions.

Atomic clocks are not stable enough to depend on just one
clock. The instability shows up as unpredictable changes in
the clock rate. A number of these clocks are used at a time in
hopes of correcting for these unpredictable changes suffered
by individual clocks. In the round-the-world flights all of their
clocks suffered changes. It is a matter of experience and judg-
ment as to how one makes time corrections for these changes.
In view of Dr. Essen's experience and analysis of all the data
one must conclude that the experiment did not prove the predicted "clock readings," much less time dilation.

12-3   Orbital motion slows a pendulum clock

The round-the-world atomic clock experiment was inconclusive. It proved nothing, In view of some later experiments,
there is a possibility that an accurate enough and stable
enough clock might have shown its running rate altered by orbital motion. Even if it did, it could be due to physical effects
altering the clock rate—a clock error—not time dilation.

One can certainly think of a reason for having the rate of a
pendulum clock (a grandfather clock) slowed by orbital motion around the earth. The period (time T for a complete tick
and tock) of a pendulum clock obeys the well known equation

$$T = 2\pi \sqrt{\frac{\ell}{g}} \qquad (12\text{-}1)$$

where ℓ is the length of the pendulum and g is the acceleration
of gravity. It is also well known that a body in orbital motion
around the earth has a centrifugal force action upward on it.
The net force, gravitation down minus centrifugal up, reduces
the effective value of g. Since g is in the denominator of Eq.
(12-1) a smaller g yields a longer duration of time T for the
completion of a tick and tock.

There is thus a physical reason why a grandfather clock would run slower in an orbital flight around the world. That, however, would be a clock error not time dilation. Time is independent of that clock. The pendulum clock's ability to measure time accurately is lost when it undergoes the centripetal acceleration that exists in orbital motion.

12-4 Ives vs. Einstein

Other scientists have suggested that instead of time being altered by motion it is the measuring device that is altered, not time. If an atomic clock is in high speed motion, physical causes may make the clock run slow, somewhat analogous to the slowing of the pendulum clock. A systematic error is introduced into the measurement of time, but time itself is not altered. To make that physical interpretation one must abandon Einstein's concepts. Herbert Ives has done the ground work in setting up one logical alternative to Einstein's special relativity.[7]

Ives showed that Einstein's treatment is inconsistent and leads to real paradoxes. He developed an alternative theory based on more experimental evidence and made a consistent case for ether as the light bearing medium. He refutes the Einstein conclusion that time is relative. He retained the common sense concept of absolute time.

Einstein claimed that time runs slower in a moving system. Ives contended there is a physical retardation in the frequency of a moving light source. This might be classified as a retardation in the clock rate, not a retardation of time itself. Both used the same equation, with source velocity v and light velocity c:

$$t = t_0\left(1 - \frac{v^2}{c^2}\right)^{1/2}$$

$$(12-2)$$

Einstein interpreted t as the time. Ives interpreted t as the clock reading, not a precise measure of the actual time t_0. Equation (12-2) gives a measure of the clock dilation, not time dilation.

Ives studied all of the famous experiments related to the relativity question. He deduced a theory that is presumably consistent with those experiments and with the ether concept. He showed that the famous Sagnac experiment (Fig. 12-1) and the Michelson-Gale experiment support an ether concept. Those two experiments employed rotating systems in which light beams were sent around a circuit in opposite directions. The phase difference observed when the two beams came to-

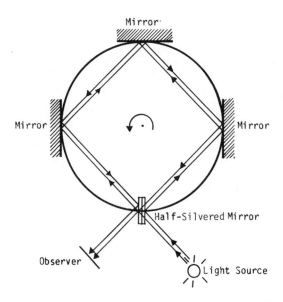

Fig. 12-1 Sagnac's experiment. The whole system, including the source of light and the observer (which is a photographic plate, not a human) is mounted on a rapidly rotating platform.

The experiment by Michelson and Gale, mentioned in the text, was similar in principle; but the apparatus was set on the Earth, which, in its rotation, served as a rotating platform. Because the Earth's rotation is slower, Michelson and Gale's experiment had to be much greater in physical extent.

gether support the concept of light having been propagated in a medium that did not rotate with the system. Ives assumed that this nonrotating reference medium is the ether. These experiments refute the Einstein concept of constancy of the speed of light. Some claimed that accelerations involved in those experiments provide a way out for Einstein, but Ives refuted that claim by showing that these arguments lead to contradictory results.

12-5   Ives-Stillwell experiment

In order to complete the case Ives and Stillwell, his colleague at the Bell Telephone Laboratories, carried out an experiment with canal rays. These canal rays consist of a beam of positively charged hydrogen molecules that emit light containing various frequencies.[8, 9] It was shown that the frequencies in this light source decrease with increase in speed of the hydrogen molecules. One would expect an observer to see a Doppler shift in the light from a moving source. However, Ives contends that his experiment shows that there is a lowering of the frequency in the light source itself, as well as the expected Doppler shift. His observed lowering of frequency in the source is given by the equation

$$f = f_0\left(1 - \frac{v^2}{c^2}\right)^{1/2} \tag{12-3}$$

where f_0 is the frequency of the source at rest. This amounts to a red shift of the center of "gravity" of the original frequencies in the light source.

The Doppler effect is the classical Doppler effect. It has a blue shift when the source is moving toward the observer. The net shift is the product of the frequency shift given by the Eq. (12-3) and the classical Doppler frequency shift. Ives interprets this as a physical effect on the light source plus the Doppler effect. Einstein gives no physical reason.

Ives' equations are similar to those of H.A. Lorentz. Ein-

stein adopted Lorentz's equations but put a different interpretation on them. Lorentz never accepted Einstein's concept
of time dilation nor his corollary concept of space contraction. Lorentz considered his equations to represent real physical effects, namely a change in clock rate and a contraction of
rod length. That is also the view held by Ives. Ives' equation
for length contraction is the same as the Fitzgerald contraction, a real physical shortening of the rod. Fitzgerald was
contemporary with Lorentz and made the original suggestion
of length contraction. (Or according to Dingle, originally
transverse expansion.) Ives was able to explain the Michelson-
Morley experiment in the same way that Lorentz and Fitzgerald explained it, with altered clock rate and altered lengths
of the arms in the instrument. The clock rate change and con-
tractions yield the null effect observed by Michelson and
Morley.

12-6 The clock paradox

Einstein's theory leads to the clock paradox. He claimed
that one clock moving at great speed would run slower than a
clock at rest. However, according to Einstein's other postulate one can not tell which clock is at rest. The result is that,
in his illustration, the clock he claimed to run fast could equally be the one to run slow, an obvious absurdity. Ives evades
that contradiction by having a standard of rest, the ether. All
motion is with respect to that fixed reference.

Ives used the same reduction ratio for both clock retardation and length contraction. That ratio is

$$\left(1 - \frac{v^2}{c^2}\right)^{-1/2} : 1 \qquad (12\text{-}4)$$

This is the same as the ratio proposed by Fitzgerald. Ives did
not claim to be able to detect the absolute value of motion
with respect to ether. He was not able to measure the total rate
at which the earth is moving through space. However, all of
his results are consistent with an ether. The clock reduction

and length contraction prevented him from detecting the value of *linear* motion with respect to the ether. That is what prevented Michelson and Morley from detecting the motion through ether. However, in the rotation experiments of Sagnac and Michelson-Gale the rotational motion with respect to the ether was measured. That is sufficient to refute Einstein's rejection of any standard of reference.

What this amounts to is that Ives has a valid theory that does not have self-contradictions, such as those in Einstein's theory. His theory deduces, in a straightforward way, a fundamental foundation for electrodynamics and checks with the basic experiments. He acknowledges the debt he owes to the early scientists by stating that his:

> views will be recognized as those of earlier students of the subject - Fitzgerald, Larmor, and Lorentz - but not of those who would shift the burden from variant measuring instruments to the nature of space and time.[10]

The last remark shows his rejection of the Einstein philosophy.

12-7   Another contradiction

V. Vergon points out the following contradictions between Einstein's special theory of relativity and the results of the Ives-Stillwell experiment. In his original paper Einstein considered the case of an observer moving with velocity v at an angle ϕ with respect to light rays from a distant star, a source considered to be at rest.[11] He derived the equation for the "observed" frequency f in terms of source frequency f_0, namely

$$f = \gamma f_0 (1 - \beta \cos \phi) \qquad (12\text{-}5)$$

where $\beta = \dfrac{v}{c}$ and $\gamma = (1 - \beta^2)^{-1/2}$

In the Ives-Stillwell experiment the source of light was moving and the observer was at rest. For simplicity, the comparison will be made for the orthogonal condition, that is to say the condition in which the angle ϕ between source or observer velocity and light rays is $90°$. Einstein's Eq. (12-5) reduces to

$$f = \gamma f_0.$$ (12-6)

The comparable Ives-Stillwell result is given by Eq. (12-3) which is recast in the form

$$f = \frac{f_0}{\gamma}$$ (12-7)

The Ives-Stillwell experiment was done with a moving source and the Einstein equation applies to a moving observer. Einstein's first postulate demands that the result for both cases be the same. That postulate states that there is no standard of rest, all motion is relative. This relative motion is equal; relativity can not distinguish which one is at rest. Einstein's relativity demands that the results of these two cases be the same. Quite obviously Eq. (12-6) and (12-7) are not the same. Equation (12-6) is an increased frequency and (12-7) is a decreased frequency. Hence, the experiment contradicts Einstein's theory.

12-8   Speed-altered spectral characteristics

Our particular interest in Ives' work is his experimental and theoretical evidence that there is a decrease in the "standard" frequencies associated with a hydrogen ion, H_2^+, when it is in high speed motion. Our explanation is:

1) That the electromagnetic fields are induced within the hydrogen as its charge components move through the reference medium.

2) That there is an interaction between the electromagnetic fields and the hydrogen's electrical components which reduces the frequency of its spectral lines.

This is consistent with the paper entitled *A Classical Foundation for Electrodynamics,*[12] which shows that movement of an elementray charge through a reference medium will, through a feedback process, develop an electromagnetic field. That field is expressed by the same equation as the relativistic field in a "preferred" frame of reference. This presumably checks with experiment. No time dilation nor space contraction is involved. That development was based on classical physics plus a reference medium and a feedback concept.

Each electrical component of the hydrogen is in the field of the other electrical components. Since those fields are altered by the movement of their associated charges through the reference medium, the interaction forces between the various charges is altered by this translational motion. The direction of these altered forces is governed by whether or not the two interacting charges are of like sign or of unlike sign. As a consequence of the two charges moving in the same direction, if the two charges are of like sign, the altered force (a magnetic force) is a pinch effect, an attraction. If the two charges are of unlike sign, the altered force is a repulsion effect. This is analogous to the attraction between two parallel currents flowing in the same direction or the repulsion force between two parallel currents flowing in opposite directions.

To explain the lowering of the spectral frequencies in the Ives-Stillwell canal ray experiment, first consider radiation of the spectral lines to be the result of vibration of the oppositely charged electrical components in the hydrogen. Motion of these charges induces a repulsion force between those components. This repulsion force is not as large as the Coulomb attraction force binding the charges together, but it does cause a weakening of that binding force. This weakening of the net force holding these components together explains the lowering of the vibration frequencies. In this case it means a lowering of the spectral frequencies, the effect that has been observed experimentally.

12-9   Increased lifetime of fast muons reinterpreted

The most frequently cited evidence of time dilation is a famous experiment in which the lifetime of fast-moving muons was greater than that of muons at rest.[13] Muons are unstable particles with an electric charge equal to that of an electron and a mass 207 times that of an electron. Fast-moving muons are produced high in the atmosphere from cosmic radiation. They can also be produced in the laboratory. The average lifetime of a muon at rest is about 2×10^{-6} seconds.

The experiment involved the statistical chance of the fast muons observed at the top of Mount Washington surviving the 3000 meter descent to sea level. The fast muons had a speed of about 0.998 of the speed of light. With no time dilation, a muon going 2.99×10^8 meter/second for $2 \times^{-6}$ second could travel only about 600 meters, not far enough to reach sea level. Whereas a statistically high percentage of muons were found to reach sea level.

The Einstein time dilation factor:

$$\gamma = \left(1 - \frac{v^2}{c^2}\right)^{-1/2}$$

turns out to be about 15 for this case. The relativist interprets this experiment to mean that time as "seen" by the fast muon ran 15 times slower than time "seen" by the experimenter. The experimenters clock gave the muon 15 times longer to reach sea level, than the muon's clock.

The author suggests another interpretation of the results of this experiment. Instead of time itself being altered by a factor of 15, with high speed motion, the stability of the muon is increased by a factor of 15. There is a physical reason why the muon does not decay so fast when it is in high speed motion. That makes a lot more sense than saying time runs differently for the muon than for the experimenter.

The physical reason for increased stability of the particle must be related to the feedback forces acting on a charged particle in motion. The feedback forces were explained in Chapters 7 and 8. These feedback forces act upon the muon in motion. One might say it is the familiar *pinch* effect, the induced magnetic attraction, that tightens up the electric charge when it is in high speed motion. It is reasonable to conclude that this "tightening up" of the particle increases its stability. Increased stability of the muon should prolong its lifetime.

A fresh look at the physical process that slows the decay rate with motion may lead to an understanding of why unstable particles decay. No one knows for example, why radioactive particles decay. It would seem that an understanding of the physical process for increasing the stability of the moving muon may shed light on what causes particles to decay. This may lead to a solution of one of the most important unsolved problems in physics, namely answering the question: *When will a particular radioactive particle decay?* That question will never be answered until physics returns to the search for physical causes.

References

1. Hafele, J.C. and R.F. Keating. 1972. Around the world atomic clocks observe relativistic time gains. *Science* 177: (4044) 168-170.
2. Essen, L. Atomic clocks coming and going. *Creation Research Society Quarterly,* Vol. 14, June 1977 p. 46.
3. November 3, 1976 letter from G. Burniston Brown to Thomas G. Barnes.
4. Essen *Op. Cit.*
5. Essen, L. 1971. The special theory of relativity, a critical analysis. Oxford Science Research papers and the Clarendon Press, Oxford.
6. Essen, L. 1972. Einstein's special theory of relativity. *Proceedings of the Royal Institution of Great Britain* 45, 141-160.
7. Turner, Dean and Richard Hazelett. The Einstein myth and the Ives papers. The Devin-Adair Co., Old Greenwich, CT. (This book gives an extensive account of Ives' work and a complete collection of his papers.)

8. Ives, Herbert E. and G.R. Stillwell. 1938. An experimental study of the rate of a moving atomic clock. *Journal of the Optical Society of America* 28 (7):215-226.

9. Ives, Herbert E. and G.R. Stillwell. 1941. An experimental study of the rate of a moving atomic clock II. *Journal of the Optical Society of America* 31 (5):369-374.

10. Ives, Herbert E., 1940. The measurement of velocity with atomic clocks. *Science* 91 (2352):79-84. See especially p. 84.

11. Vergon V. 1976. Relativity beyond Einstein. Exeter Publishing Co., Los Angeles.

12. Barnes, Thomas G., Richard R. Pemper, and Harold Armstrong. 1977. A classical foundation for electrodynamics. *Creation Research Society Quarterly* 14 (1):38-45.

13. Frisch, D.H. and J.H. Smith. *American Journal of Physics* 31, 342 (1963).

CHAPTER 13

The Neutron and Nuclear Forces

13-1   What holds the nucleus together?

One of the mysteries of physics has been the nature of the force that holds the nucleus together. The nucleus of an atom always has a net positive charge. It contains protons and protons have a positive charge. The only other particle in a nucleus is the neutron and, as the name implies, it has no net electric charge. The question is: Why don't the protons in the nucleus fly apart? Like charges repel and Coulomb's law tells us that the electrostatic force between two protons that close together would be very strong.

There obviously has to be some very strong force within the nucleus holding those protons within the nucleus. That force, whatever it is, is called the *strong force*. It is a short range force that is not noticeable outside of the nucleus. So it is also known as a *nuclear force*. Modern physics looks upon this nuclear force as something quite different from electric, magnetic, or gravitational force.

The author proposed in his previously published paper, New Proton and Neutron Models, that this strong force within the nucleus is a magnetic force.[1] Never before in physics had it been thought that the magnetic force between charged particles could exceed the electrostatic force between those particles. That would still be true if one blindly followed the Einstein concept that under no condition can the motion of electric charge exceed the speed of light. We have seen in Sec. 10-11 that the surface of the rotating proton can exceed the speed of light. We shall show that this can yield a larger magnetic force than electric force. That phenomenon makes it possible for the strong force within the nucleus to be a classical magnetic force.

13-2   Rotational speeds exceeding the speed of light

The previously mentioned feedback theory was developed as a classical alternative to special theory of relativity.[2] An extension of that theory to a rotating charged body, such as a proton with rim speed exceeding the speed of light c, opened the way for logical applications to atomic and nuclear physics. The limitation of velocity to the speed of light, according to this theory, results from the feedback effect on a finite charged particle in translational motion. As a finite particle moves along, it produces a changing magnetic field at a fixed point in the medium. This changing magnetic field induces an electric field that is fed back. This feedback enhances both the electric field and the magnetic field of the moving charge. This feedback enhancement causes the E and B fields to approach infinite magnitude, putting a limit on the translation speed of a charged particle.

If there is no fluctuation in the B field, as for example in uniform rotational motion of a spherical charged particle, there is no feedback. With no feedback there is no increase in the electric field, only the nonrelativistic increase in the magnetic field. Under this no-feedback condition, it is possible to exceed the speed c and to achieve an increase in the ratio of magnetic to electric field.

The increase in the magnetic field H is brought about by the fundamental relation

$$\mathbf{H} = \mathbf{v} \times \mathbf{D} \qquad (13\text{-}1)$$

From that equation and the relation $B = \mu H$, and the Maxwell equation for the speed of light,

$$c = (\mu\varepsilon)^{-1/2}$$

one may deduce the equation

$$\mathbf{B} = \frac{\mathbf{v}\times\mathbf{E}}{c^2} \tag{13-2}$$

In the right angle case and where a second charge q_0 is moving with speed v_0 with respect to this B field, the magnitude of the force on that charge is

$$\mathbf{F} = \frac{q_0 v_0 v E}{c^2} \tag{13-3}$$

In the case of a positively charged sphere spinning with uniform angular speed there is no rate of change of B. The rotational speed v of its rim can exceed the speed of light c. Equation (13-2) still applies to the B field. When v increases B increases but E remains constant. When v exceeds c there is a greater ratio of B to E than would be possible in the feedback case. Now consider a negative charge moving with speed v_0 through this stronger magnetic field. It is possible for the magnetic force to exceed the electric force on that charge, repelling it instead of attracting it. That relative increase of magnetic to electric force is, of course, also a function of the speed v_0. The greater v_0 the greater that ratio of magnetic to electric force. Here then is a possible mechanism to achieve a large enough magnetic field to hold apart a spinning proton and a revolving electron.

13-3 Magnetic barrier between electron and proton

From Eqs. (10-18) and (10-19) and the value of magnetic moment, charge, and radius of the proton, the angular velocity $\omega = 2.52 \times 10^{29}$ rad/sec and the rim speed $v = 2.58 \times 10^{11}$ meter/sec. That high speed rotation produces an exceedingly strong magnetic field near the proton, a magnetic barrier that the rotating electron can not penetrate.

The positive charge of the proton attracts the electron but the proton's magnetic field exerts a repulsion force on the rotating electron. Of particular interest is the relative value of the magnetic repulsion pressure to the electric attraction tension on the nearby surfaces of the spinning electron and the spinning proton in our model of the hydrogen atom. As an elementary illustration of that pressure barrier provided by the strong magnetic field between the spinning proton and electron consider the magnetic force dF_m on an elementary surface charge dq of the electron near the rim of the proton. Assume that the electron's surface has the speed given by Eq. (10-19) with $M = 9.285 \times 10^{-24}$ and $r = 1.87 \times 10^{-15}$. The electron's surface speed $v_0 = 9.30 \times 10^{10}$ meter/sec.

Substituting the differential element notation into Eq. (13-3) the magnetic repulsion force on that small adjacent portion of the electron is

$$dF_m = \frac{dq v_0 vE}{c^2} \qquad (13\text{-}4)$$

The associated Coulomb attraction force is

$$dF_e = dq\, E$$

$$(13\text{-}5)$$

Substituting for the proton rim speed in terms of the speed of light c, $v = 861c$ and the electron rim speed $v_0 = 310c$, one has the ratio of magnetic repulsion pressure to electric attraction tension on this surface of the electron,

$$\frac{dF_m}{dF_e} = 267,000 \qquad (13\text{-}6)$$

The surface of the electron will not come into contact with the surface of the proton. The positive and negative charges will remain separated. The *net* electric force will, however, pull the two into where they will form a depression in the electron as shown in Fig. 11-3.

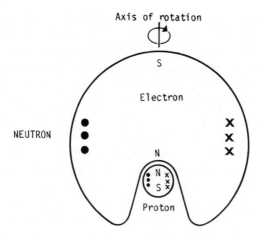

a) Section through the depression

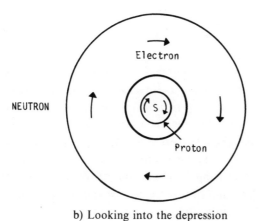

b) Looking into the depression

Fig. 13-1 Neutron consisting of a spinning proton and a spinning electron, both spinning in the same direction. The electric attraction between the positive charge of the proton and the negative charge of the electron is balanced by the magnetic repulsion of these two magnets.

13-4   Neutron model

We have already shown that a rotating electron satisfies the non-radiating condition for a classical model of the hydrogen atom. As previously mentioned, this is superior to the orbital electron model in the Bohr atom, the defect in the orbital model being loss of energy through radiation. In the stable state of the hydrogen atom the magnetic moments of the adjacent spinning proton and rotating electron are parallel, pointed in the same direction, as shown in Fig. 11-4. They spin in opposite directions.

Our present interest is in an application of the electron and the associated proton sphere to the new model of the neutron. Even though the neutron has no net charge it has a magnetic moment. That implies a rotation of charges. Hence, it is logical to assume that it contains an electron and proton in some type of rotational motion. As was true in the hydrogen atom the electron and proton must be held apart by a magnetic force, so that they cannot discharge.

Assume that the electron in Fig. 11-4 is forceably flipped over 180 degrees and pushed into the equilibrium position with respect to the proton to form a neutron as shown in Fig. 13-1. The direction of rotation is the same for the electron and the proton. The electrostatic attraction and magnetic repulsion achieve a balance condition. The adjacent portions of the electron and proton currents have opposite directions, providing a strong magnetic repulsion to prevent the merging of the plus and minus charges. Because one charge is plus and the other minus, their magnetic moments have opposite directions.

The non-radiating condition holds for the electron and proton in the neutron for the same reason it holds in the hydrogen atom. This combination of the electron and proton satisfies the following requirements for a neutron: 1) no net charge, 2) extremely small size, 3) has a magnetic moment, 4) has less stability in the free state than the stability of an electron and proton in the hydrogen atom, and 5) releases radiation energy and kinetic energy when the neutron decays into an electron and proton.

13-5   Forming a neutron

Let us illustrate a hypothetical means of forming a neutron. It requires the addition of external energy and constraints that orient the proton and electron in the desired directions until the neutron is formed. The proton has its intrinsic spin ω_0 at the beginning. The electron has no spin until a spin is induced from the proton's magnetic field as the electron is moved in toward the proton.

The formation process is illustrated in Fig. 13-2. The electron is moved in on a straight line toward the north pole of the spinning proton. The electron's induced spin is such as to produce a magnet with its north pole pointing toward the proton. The applied constraints prevent the proton or electron from orienting themselves out of this magnetically repulsive orientation. The electrostatic attraction is only a part of the force moving these two together. Additional force is added to bring the electron into this stronger magnetic field, with the north pole opposite north pole. Finally, the proton locates in the indentation in the electron forming the neutron as shown in Fig. 13-1. This is an unstable equilibrium condition with enough energy for the neutron to decay if it is not in an atom where the combination of forces stabilize it.

The magnetic moments M_e and M_p of the electron and proton in the neutron point in opposite directions. The electron's magnetic moment is designated as minus since it points in the opposite direction from the spin angular momentum. Note that the electron and proton spin in the same direction and about the same axis (Fig. 13-1).

The energy put into the system will be shared by the electron and proton. The proton will gain the larger fraction of

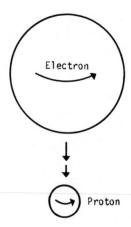

Fig. 13-2 Formation of a neutron. External force moves the electron
toward the proton such that their magnetic dipoles point toward each other. The magnetically induced spins are in the
same direction.

the energy. That is due to the much smaller moment of inertia of the proton. The induction that takes place causes the
electron to spin faster to oppose this increase in magnetic
flux. Its own magnetic flux links the proton, whose spin also
increases to oppose that change in flux. The magnetic repulsion between the two increases until it balances out the electrostatic attraction.

Experiments have presumably shown that the magnetic
moment of the neutron $M_n = -5.05 \times 10^{-27}$ amp m^2 and has
the same direction as the magnetic moment M_e of the electron. One of the design requirements for the neutron is this
equation relating the magnetic moment,

$$M_n = M_e + M_p \qquad (13\text{-}7)$$

The value of the magnetic moment for the proton in the
hydrogen atom is

$$M_p = 1.41 \times 10^{-26}$$

and for the electron is

$$M_e = 9.285 \times 10^{-24}$$

One can see from Eq. (13-7) that the spin and associated magnetic moment of the proton has to be very much larger in the neutron than it is in the hydrogen atom. That is one of the reasons why the neutron has enough "excess" energy to produce the beta decay and all of its related energy components. That is also one of the reasons why it is difficult to form a neutron from an electron and a proton.

13-6 Beta decay and the neutrino reinterpreted

It is our contention that modern physics has trapped itself into an awkward position by insisting on the quantum spin constraint and an erroneous notion about the equivalence of mass and energy. A good example of that awkward position can be found in the conventional modern physics interpretation of *beta decay* of the neutron. The term beta is associated with an electron in high speed motion. Modern physics denies that there is an electron in the neutron, even though the neutron decays into a proton and an electron in high speed motion.

A well known modern physics textbook points out that beta decay presents a rather difficult problem to the physicist who seeks to understand natural phenomena:

The most obvious difficulty is that in beta decay a nucleus emits an electron, while, as we have seen in the previous chapter, there are strong arguments against the presence of electrons in nuclei. Since beta decay is essentially the spontaneous conversion of a nuclear neutron into a proton and electron, this difficulty is disposed of if we simply assume that the electron leaves the nucleus immediately after its *creation*. [emphasis added] A more serious problem is that observations of beta decay reveal that three conservation principles, those of energy, momentum, and angular momentum, are apparently being violated. . .

The energies observed in the beta decay of a particular nuclide are found to vary continuously from 0 to a maximum value T_{max} characteristic of the nuclide. . . . In every case the maximum energy

$$E_{max} = m_0 c^2 + T_{max}$$

is equal to the energy equivalent of the mass difference between the parent and daughter nuclei. Only seldom, however, is an emitted electron found with an energy of T_{max}.[3]

To correct for the "missing energy" and the momentum problem in the decay process, Pauli proposed the emission of an uncharged particle of zero mass and spin $1/2$, later called *neutrino*. The neutrino was supposed to have energy equal to the difference between the T_{max} energy and the actual electron kinetic energy and to have the "missing momentum."

Our neutron has been developed from the classical principles of conservation of energy and momentum, both linear and angular momentum. One does not have to "create" an electron nor invent a new massless particle, a neutrino. Everything has been taken care of by classical physics in our model.

Even though our model of the neutron is by no means complete and quantitative at the present stage of development, its eventual development must be governed by the above mentioned conservation principles. The basic concepts are these: inertial mass of the neutron depends upon the size and configuration of the electron and proton in the neutron configuration. Energy is conserved in every process associated with the neutron. In all of the self-acting processes momentum is conserved, both linear and rotational momentum. The magnetic energy associated with the spins must be taken into account in the conservation of energy. Electromagnetic induction causes the changes in the spin and associated magnetic moments of the electron and the proton.

When a neutron decays it releases its proton and electron and the conservation of momentum, both linear and rotational, is continually maintained by these electrical components. Some of the neutron's energy is released in the form of kinetic energy delivered to the electron and proton. The excess energy is released as electromagnetic radiation. That radiated energy takes the place of the modern physics neutrino.

13-7   A symmetrical neutron model

In view of the tentative status of the neutron model presented in the previous sections, a second model is mentioned as an alternate possibility. This alternate model has more symmetry and wider range of stability, while still meeting the condition of instability where needed. This model is similar to the electron ring model the author developed in a previous paper.[4]

Figure 13-3 illustrates this alternate model of the neutron. It consists of the spinning proton located in the center of a hole through the middle of the spinning electron. Figure 13-3a is a view of a section through the middle parallel to the axis of the hole. Figure 13-3b is a view looking down into the hole. Note that the electron and proton are spinning in the same direction and about the same axis, the axis through the center of the hole.

The magnetic field generated by the very rapidly rotating proton provides the strong magnetic repulsion force on the rotating electron to keep the electron charge separated from the proton charge. At the midpoint on the axis both the magnetic force and the electrostatic force on the proton are zero, providing an equilibrium condition.

The spin of the proton in the neutron is much greater than the spin of the proton in the hydrogen atom. Excess energy is needed to form the neutron and it has gone primarily into

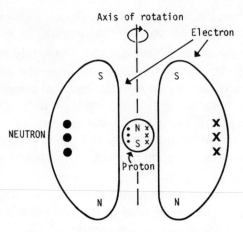

a) Section through the hole

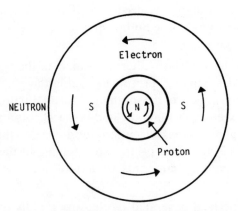

b) Looking into the top of the hole

Fig. 13-3 Symmetrical neutron model.
The neutron consists of a spinning proton and electron. The electron has a hole through its axis of spin. The proton is located inside of the hole at the center. They spin about the same axis and in the same direction. The electron charge and proton charge are held apart by the magnetic field.

the spin energy of the proton. Some of this excess energy will, in the decay of the neutron, be transformed into electromagnetic radiation energy, the energy that modern physics has assumed to be a neutrino. The rest of it will provide the kinetic energy of the ejected electron when the neutron decays. The momentum is conserved in the electron-proton combination in accordance with classical physics. There is no need for "creation" of a spin in a neutrino.

Figure 13-4 illustrates the inherent instability of the neutron once the proton is located outside of the electron hole as shown. Note that if the proton is above the electron, as in Fig. 13-4a, its south pole repels the south pole of the electron, ejecting the lighter electron downward. Once the proton is located down outside of the electron hole, as in Fig. 13-4b, the north pole of the proton repels the north pole of the electron, causing the electron to be ejected upward. This instability is a part of the beta decay mechanism.

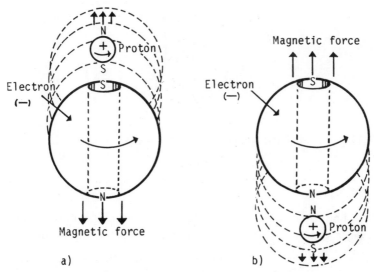

Fig. 13-4 Neutron decay.
Two unstable conditions of the neutron illustrating the dominant force on the electron during beta decay.

13-8   Proton-proton force balance

We have seen that the electrostatic attraction between the electron and proton can be balanced in the neutron and in the hydrogen atom by the magnetic forces resulting from their high spin rates. The next problem is to explain the forces that hold two protons within the nucleus. The electrostatic repulsion force on these two positive charges tends to force them out of the nucleus.

Each proton acts initially like a permanent magnet because of their intrinsic spin. That means that they tend to be drawn together with the north pole of one attracting the north pole of the other as shown in Fig. 13-5.

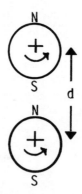

Fig. 13-5 Magnetic attraction between protons.

The equation for the magnetic attraction force of the magnetic dipole of one proton on the magnetic dipole of the other proton is

$$F_m = -\frac{3\mu\ M_p^2}{2\pi\ d^4} \tag{13-8}$$

The Coulomb electrostatic repulsion force is

$$F_e = \frac{q^2}{4\pi\varepsilon d^2} \qquad (13-9)$$

It is easy to solve for the separation distance d_0 where these oppositely directed forces are equal.

If there were no other factor involved this would be a very *unstable* balance point. The electrostatic repulsion force is greater at distances larger than d_0. The magnetic attraction force is greater at distances smaller than d_0.

The instability is due to our initial assumption that these protons have permanent magnets because of their intrinsic spin. However, we have previously pointed out that the proton is also subject to Faraday's induction law. Its spin is altered whenever external magnetic flux links it.

This magnetic induction between the protons provides the additional factor that prevents the protons from crashing on into each other once they get closer together than that unstable distance d_0. By Lenz's law the induced spin is such as to oppose that motion. In other words, the proton spin slows down as these two magnets approach each other, reducing the excessive attraction.

Classical physics can provide a reasonable explanation of the "nuclear" forces that hold two protons within the nucleus. It will take years, no doubt, for scientists to work out the precise relations but it appears that the classical approach can provide the answers.

References

1. Barnes, Thomas G. New proton and neutron models. *Creation Research Society Quarterly,* Vol. 17, June 1980, pp. 41-47.
2. Barnes, Thomas G., Richard R. Pemper, and Harold L. Arm-

strong. 1977. A classical foundation for electrodynamics. *Creation Research Society Quarterly* 14(1):38-45.
3. Beiser, Arthur. Concepts of Modern Physics, Second Ed., McGraw-Hill, New York, 1973, p. 408.
4. Barnes, *Op. Cit.*, 1980.

CHAPTER 14

Atomic Radiation

14-1 Hydrogen spectral frequencies

Bohr was able to predict the spectral frequencies of the hydrogen atom from the equation

$$f = Rc \left(\frac{1}{n_f^2} - \frac{1}{n_i^2} \right) \tag{14-1}$$

in which the Rydberg number $R = 1.0974 \times 10^7$, and $c = 2.9979 \times 10^8$ (speed of light). the n_i and n_f are integers and may have any value so long as $n_i > n_f$.

According to Bohr's theory of the hydrogen atom the various spectral frequencies are radiated when the orbital electron falls from a higher state, represented by n_i, to a lower state represented by n_f. The *ground state* is represented by $n_f = 1$. The atom receives energy (is excited) by some electric or magnetic process and is raised to a higher state and may fall back through successive stages radiating a spectral frequency for each step in accordance with Eq. (14-1).

For example, in a typical case, it may be raised to state $n = 7$ and fall back successively through states $n = 4$ and $n = 2$ to the ground state $n = 1$. Three spectral frequencies will be emitted in accordance with Eq. (14-1): for $n_i = 7$ and $n_f = 4$, $n_i = 4$ and $n_f = 2$, and $n_i = 2$ and $n_f = 1$. These spectral frequencies are respectively $f_1 = 1.3827 \times 10^{14}$ Hz, $f_2 = 6.1592 \times 10^{14}$ Hz, and $f_3 = 2.4637 \times 10^{15}$ Hz.

The complete spectrum includes all the possible transitions that occur where there are a very large number of excitation and de-excitation processes taking place. That is a condition that might be produced by many collisions during an electric discharge.

The *intensity* of these hydrogen spectral lines varies greatly. A few of them are very bright. However, many of them are so dim, or dim and close together, that they are difficult to detect. Bohr's theory gives no clues to the intensity of the spectral lines.

It is possible to enhance the intensity of one of the dominant spectral lines. A *laser* can enhance the light emitted from a dominant spectral line, to such an extent that the other lines have negligible intensity. In a laser the light is reflected back and forth between two parallel mirrors. The resultant *standing wave* of light causes the atoms to be *excited* more and more in the dominant mode.

Whereas ordinary light is incoherent, the light emitted from a laser is coherent. The standing electromagnetic wave in the laser causes the excitation to take place in such a synchronized phase that the resultant light emitted from the atoms is coherent. This coherence makes it possible to beam that light much more sharply. Surely there is a model and a classical physical cause for the spectral radiation and laser in-phase oscillations taking place in the hydrogen atoms. It does not seem very scientific for modern physics to have abandoned the search for a classical explanation of the hydrogen atom and its radiation. There must be physical causes for all of these physical effects.

14-2 Radiation mechanism

Quantum theory provides neither a physical model nor a radiation mechanism for the atom. There is a well known classical mechanism for electromagnetic radiation. It is acceleration of charge. The vibratory acceleration of charge in an antenna produces electromagnetic radiation such as that in radio and radar.

Any time an electron or a proton is accelerated there is radiation. There is no physical reason for abandoning this principle in atomic physics. There is a justifiable complaint that no one has come up with an acceptable classical model of the atom and that no one has come up with a classical means of generating the vibratory accelerations of the charges within the atom to yield the known spectral frequencies.

The model of the hydrogen atom presented in Chapter 11 is not yet fully developed. However, it does demonstrate that there can be new approaches to this basic problem. It is regrettable that so little scientific effort has been applied to this very basic problem. One must lay the blame for that on the philosophical indoctrination that has accompanied the promotion of the relativistic and quantum mechanical view point. Looking at this in another way, one may say that here is a great opportunity for a young scientist of this day to make some very fundamental contributions, without much competition.

In Bohr's theory the quantum of energy

$$w = h\nu \qquad (14\text{-}2)$$

is supposed to be emitted when an electron makes a "quantum jump" from an orbit of larger radius to one of smaller radius. A single jump downward in orbital radius would not produce a coherent electromagnetic wave train of a single frequency. Changing the notation for frequency from ν to f,

$$w = hf \qquad (14\text{-}3)$$

In Sec. 6-4 the conclusions of H.A. Lorentz and Herbert Ives, based on extensive optical experimental evidence, indicate that the emission from each individual atom consists of many waves that are coherent over distances greater than one meter. A high Q resonant vibration of an electric charge could certainly produce that effect. It is believed that when a physical model of the hydrogen atom is finally perfected it will have the inherent modes to yield the observed frequencies of the hydrogen atomic spectrum.

According to the Bohr theory the highest possible spectral frequency for the hydrogen atom would be emitted if the electron fell in from infinite orbit, $n_i = \infty$, to ground state orbit $n_f = 1$. From Eq. (14-1) the highest frequency $f_{max} = Rc = 3.299 \times 10^{15}$Hz. According to Eq. (14-3) the maximum quantum of energy emitted by hydrogen

$$w_{max} = hf_{max} = 2.180 \times 10^{-18} \text{joule.} \quad (14\text{-}4)$$

This value of maximum energy for the hydrogen atom is consistent with the *ionization potential* of the hydrogen atom. Recalling that the energy required to raise a charge q to potential V is given by

$$\text{Energy} = qV \quad (14\text{-}5)$$

and substituting the value of the charge of an electron, $q = 1.602 \times 10^{-19}$ coulomb, and the above value of energy into Eq. (14-5) yields the *ionization* potential

$$V_H = 13.6 \text{ volts}$$

for the hydrogen atom. That is the experimentally verified value of ionization potential for the hydrogen atom.

One does not need quantum theory to deduce the energy value in Eq. (14-4). The energy required to ionize the atom equals the radiation energy emitted by the vibrations before the electron settles down, when it has fallen back into the atom. That follows from the law of conservation of energy.

One thing should be noted. The intensity of the spectral line is not greatest for the largest quantum of energy. The intensity depends upon the superposition of the radiation from many atoms. It is not a measure of any quantum of energy. Following the lead of Ives and Lorentz, we reject the photon concept. Light is considered to be a wave, not a particle. From our point of view the intensity of a spectral line is due to the radiation from many atoms yielding the kind of field pattern electrical engineers would predict classically.

The directional and enhanced emission of light in a laser appears to be something that one would expect classically. The energy that is reflected back and forth in the standing wave pattern should be expected to yield in-phase excitation and emission that produces the laser beam, very much like a radar beam is produced by the phasing in an antenna. The beam pattern is predicted by wave interference encountered by the emissions from all of the atoms.

One might also look at laser phenomena somewhat like oscillations produced by an amplifier with positive feedback in the axial direction of the laser. The interpretation of a laser should not be restricted to quantum theory. There is a need, however, to develop a classical model of the atom which can have the required vibrational modes.

14-3   High Q atomic oscillators

A reasonable explanation of spectral radiation from the hydrogen atom is vibrational modes of the electric charge system within the atom. These vibrations might be vibrations of: 1) the electron as an entity, 2) the proton as an entity, 3) structural vibrations within the electron itself, or 4) some combination of these vibrations. There are an almost endless number of possible vibrational modes in the atomic model described in Chapter 11 and illustrated in Fig. 11-3.

A unique feature of these atomic vibrational modes is that there is *no ohmic loss,* no ohmic resistance. The only loss of energy is in the radiation. Even there the so-called *radiation resistance* is extremely small. This means that this vibratory system has an extremely high Q. That means it has very sharp resonances and can freely vibrate (once excited) for an extremely large number of cycles.

The sharp resonances mean that the dominant frequencies are sharply defined. The high Q property provides the kind of mechanism that when excited into vibration can continue to vibrate freely for a relatively long period. The radiated waves from one atom are coherent throughout a distance of more than a meter. That is consistent with the experimental observations referred to by Lorentz and Ives, of light waves that showed coherence through a distance of more than two million wave lengths.[1]

14-4 Resonant inductances and capacitances

For each fundamental resonance in the spectral radiation of an atom there is an equivalent inductance L and capacitance C in accordance with the well known equation for resonant frequency

$$f = \frac{1}{2\pi \sqrt{LC}}$$ (14-6)

We shall illustrate the method of deducing these magnetic and electric parameters for the hydrogen alpha spectral frequency

$$f_\alpha = 4.56803 \times 10^{14} Hz.$$

We shall consider this vibrational frequency to be analogous to that of plucking a violin string, where a certain amount of added potential energy is the excitation energy. Denote that excitation potential energy as $W\alpha$. We assume the energy value

$$W_\alpha = 3.02685 \times 10^{-19} joule$$

is this excitation potential energy. This is equivalent to 1.88918 electron volts, the same energy that one would compute from the quantum equation $W_\alpha = hf_\alpha$. Whereas in quantum theory the system could only vibrate at that frequency when *exactly* that amount of energy is added; in our classical theory this is the *maximum* amount of excitation that will yield that resonant frequency.

In a high Q resonant system the energy oscillates back and forth from electric energy to magnetic energy. In the initial oscillations one may equate the excitation energy to the *peak* electric or *peak* magnetic energy. Equating this peak magnetic energy to the excitation potential energy

$$\frac{LI_0^2}{2} = W_\alpha \qquad (14\text{-}7)$$

where I_0 is the peak electric current in this vibrational motion. The peak current

$$I_0 = \sqrt{2}\ f_\alpha q \qquad (14\text{-}8)$$

where fq is the effective current in this sinusoidal vibration and the multiplication by the square root of two converts it to peak value of current. Substituting Eq. (14-8) into (14-7) and solving for inductance

$$L = \frac{W_\alpha}{f_\alpha^2 q^2} \qquad (14\text{-}9)$$

Evaluating from these known values yields the resonant inductance for this spectral vibration

$$L = 5.6507 \times 10^{-11} \text{henry}.$$

All values are now known in Eq. (14-6) except C. Solving that equation for C and evaluating yields

$$C = 2.1423 \times 10^{-21} \text{ farad}.$$

14-5 Radiation resistance

The atomic radiator is a *perfect* radiator in the sense that it has no internal resistance heat loss. The average power P_{av} radiated may be expressed in terms of effective current I = qf and radiation resistance R

$$P_{av} = q^2 f^2 R$$

and also as

$$P_{av} = \frac{W_\alpha}{T}$$

where W_α is the excitation energy, and T is the time constant of the freely decaying oscillations. From those equations it follows that

$$R = \frac{W_\alpha}{f^2 q^2 T} \qquad (14-10)$$

The value of the time constant T is not yet known. We shall assume that $T = 10^{-8}$ sec, which is the time frequently suggested in modern physics as the time an atom stays in an excited state.[2] Substituting that value and the known values into this equation yields the radiation resistance value

$$R = 5.65067 \times 10^{-3} \text{ohm}.$$

As a check on the consistency, one may substitute that value into the decay time constant equation

$$T = L/R$$

and get the expected value $T = 10^{-8}$ sec. It should be pointed out that this value of radiation resistance depends upon the *assumed* value of the decay time. A more accurate value will eventually have to be measured by some experimental means. Nevertheless, this value is good enough to illustrate the classical theory.

14-6 Effective mass in atomic radiation

One of the difficulties in developing a classical model of atomic vibrations to fit the known radiation spectra is related to the effective mass. The spectral frequencies are lower than the resonant frequencies one would obtain in an equivalent mechanical system of a model such as that of the Bohr atom. For one thing the mass of the electron and of the proton is too small. It takes a *larger* mass associated with the model's *stiffness* to yield the required lower frequencies.

The classical model we have proposed has a closely

*coupled magnetic system in which the effective mass is large
enough to yield the required lower resonant frequencies.*
The following derivation illustrates why the *effective mass*
of the atom's vibratory system is much larger than the mass
of the electron or proton. Engineers familiar with electro-
mechanical transducers, such as the loud speaker in a radio,
are familiar with the fact that the effective mass in a radiating element is greater than the mass of that vibrating element.

Recalling that the kinetic energy of a moving charge is
equivalent to its magnetic energy, one may express that
equation as

$$1/2 \ mv^2 \ = \ 1/2 \ LI^2 \tag{14-11}$$

where m and L are the *effective mass* and *effective inductance* in the vibratory system. The v and I are the speed
and current in the vibratory motion of the charge.

Because the motion is sinusoidal, the peak speed v_0 is related to the peak displacement s_0 by the equation

$$v_0 \ = \ \omega s_0 \tag{14-12}$$

where $\omega = 2\pi f$. In view of Eqs. (14-11) and (14-12) the equation for effective mass may be written

$$m \ = \ \frac{L \ I_0^2}{\omega^2 s_0^2} \tag{14-13}$$

By aid of Eq. (14-9) and $I_0 = \sqrt{2} \ fq$ the effective mass equation becomes

$$m \ = \ \frac{2W_0}{\omega^2 s_0^2} \tag{14-14}$$

where W_0 refers to the initial excitation energy and s_0 the
maximum displacement in the oscillatory motion.

This effective mass is much larger than the mass of an electron or proton. The reason for this increase in effective mass is the inertial-like resistance to vibratory motion due to the Lenz's law effect of the induced magnetic energy in this magnetically coupled system. This inertial-like reaction upon the moving charge increases its effective mass. The quantity W_0 may be thought of as the oscillatory magnetic energy induced into the system. It is equal to and derived from the excitation energy.

To get some notion as to the plausible increase in the magnitude of the effective mass, let us assume that this resonant frequency vibration takes place with a maximum displacement $s_0 = 1.5 \times 10^{-15}$ meter, as a result of the addition of excitation energy $W_0 = 3.027 \times 10^{-19}$ joule. Using those values and the known frequency $f_\alpha = 4.568 \times 10^{14}$ Hz in Eq. (14-14) yields an effective mass

$$m = 3.27 \times 10^{-20} \text{kg}$$

Note that this is much larger than the mass of the electron

$$m = 9.1 \times 10^{-31} \text{kg},$$

or of the proton,

$$m = 1.67 \times 10^{-27} \text{kg}.$$

Since the displacement s_0 associated with excitation energy is not known, this value of effective mass is not the true value but it shows the plausibility of an effective mass being much larger than the mass of an electron or proton.

Equation (14-14) gives a clue to one possible structural configuration of our classical model of the hydrogen atom. The quantum constraint of only one fixed value of radiation energy at this frequency has been rejected. We must show that the structure can respond to a smaller initial excitation energy, radiating a smaller than quantum amount of energy. To maintain a constant effective mass at that resonant frequency, one might assume that the structure has such a configuration that the lesser W_0 is proportional to the initial displacement s_0^2, over the whole range in which it vibrates at that frequency. Perhaps that consideration can guide one to a more complete specification of our model of the hydrogen atom.

Let us show the plausibility of a model configuration that meets that requirement, namely of having the excitation energy W_0 proportional to s_0^2, to the square of displacement. The potential energy of two spherical charges q_1 and q_2 whose centers are separated distance r is

$$\text{P.E.} = \frac{q_1 q_2}{4\pi \epsilon r} \tag{14-15}$$

Let charge q_1 be that of our very small spinning proton. Assume that the proton moves on into the electron, not touching it, but moving into the indentation such that it is essentially *inside* the electron. That outer portion of the electron's charge is no longer effective. The effective charge of the electron

$$q_2 = \frac{4}{3}\pi r^3 \rho$$

where ρ is the charge density and assumed to be uniform. Eq. (14-15) for potential energy reduces to

$$\text{P.E.} = \frac{q_1 \rho r^2}{3\epsilon} \tag{14-16}$$

meeting that condition of maintaining a constant effective mass in this resonant system for all amplitudes of vibration within that structural range.

14-7   Resonant vibrations

A mechanical system with a restoring force

$$F = -ks \tag{14-17}$$

will resonate at the frequency

$$f = \frac{1}{2\pi} \sqrt{\frac{k}{m}} \tag{14-18}$$

where k is the stiffness and m is the effective mass. The plausibility of that type of force in the hydrogen atom can be shown as follows. The net force acting on the electron is the result of a magnetic repulsion and an electric attraction. Assume that the magnetic force is that between two magnetic dipoles separated distance r and oriented as shown in Fig. 11-4. The magnetic force

$$F_m = \frac{3\mu \, M_p \, M_e}{4\pi \, r^4} \tag{14-19}$$

where M_p and M_e are the magnetic moment of the proton and electron.

The electric force

$$F_e = \frac{-q^2}{4\pi \varepsilon r^2} \tag{14-20}$$

The net force is the algebraic sum of those two forces.

Let us denote the separation distance as r_0 where the net force is zero, the stable position. The vibrations will take place about that location. The restoring force will tend to return the electron to that position.

Replacing r by $r_0 + \Delta r$ the net force may be written as

$$F = \frac{3\mu \, M_p \, M_e}{4\pi (r_0 + \Delta r)^4} - \frac{q^2}{4\pi \varepsilon (r_0 + \Delta r)^2} \tag{14-21}$$

Making the approximation that the displacement Δr is negligible with respect to r_0, the equation can be reduced to

$$F = \frac{-q^2 \, \Delta r}{2\pi \varepsilon r_0^3} \tag{14-22}$$

The remainder of the equation was equal to zero, representing the condition of balance.

Equation (14-22) is of the same form as Eq. (14-17). The stiffness constant is

$$k = \frac{q^2}{2\pi\epsilon r_0^3}$$

(14-23)

and the displacement s is $\triangle r$. Substituting this stiffness into Eq. (14-18) this atomic resonant frequency

$$f = \frac{1}{2\pi}\sqrt{\frac{q^2}{2\pi\epsilon \, mr^3}}$$

(14-24)

The equilibrium position r_0 can be found by equating Eqs. (14-19) and (14-20) for that value of separation. That equation is

$$r_0 = \sqrt{\frac{3\mu\epsilon \, M_p M_e}{q^2}}$$

(14-25)

Using $M_p = 1.41 \times 10^{-26}$ and $M_e = 9.28 \times 10^{-24}$ yields the equilibrium separation distance $r_0 = 1.3 \times 10^{-14}$ meter.

That value of r_0 is too large. As seen in Sec. 10-8 the electron does not acquire that large a magnetic moment until it is more closely coupled to the magnetic flux of the proton. Hence the final separation distance will be much smaller, as can be seen in Fig. 11-3.

The need for close magnetic coupling to provide the required large effective mass means that r_0 should be smaller than the radius of the electron. For illustrative purposes let us assume that $r_0 = 10^{-16}$ meter and that the resonant frequency $f = 4.67 \times 10^{14}$. Equation (14-24) yields an effective mass

$$m = 5.36 \times 10^{-11} \text{kg}$$

This shows that the effective mass of the atomic system must be vastly larger than the electron mass for the system to resonate at the spectral frequencies. The close magnetic coupling in this model of the hydrogen atom can yield sufficient *effec-*

tive mass in the resonant system to provide the required spectral frequency.

14-8   Additional spectral lines

There is a means in circuit theory of obtaining two resonant frequencies when a coupled tuned circuit is tuned to one frequency. That is to say when the inductance and capacitance are tuned to one frequency there will be two resonant frequencies, neither of which is that frequency to which the circuit elements were tuned. That type of double resonant frequency exists in a radio circuit in which the transformer coupling is closely coupled.[3] It is used in radio alignment to broaden the frequency response.

The magnetic coupling in our model of the hydrogen atom reminds one of that type of transformer coupling in a tuned radio circuit. There should be two resonant peaks produced under certain coupling conditions, not just one resonant peak at the "expected" frequency. The amount of frequency separation is governed by the "tightness" of the coupling. The sharpness of the peaks is still governed by the Q of the system. Since there is no ohmic resistance in the hydrogen atom the Q of the system will be very high, allowing for sharp lines.

The two resonant peaks lie on opposite sides of the frequency to which the system is tuned, one at a higher frequency and the other at a lower frequency. This reminds one of *line splitting* in the fine spectra of an atom. One line is at a higher frequency and the other at a lower frequency than, for example, the expected Bohr frequency. This magnetic coupling in the hydrogen atom may be the mechanism for producing some of the spectral lines predicted by Bohr and/or the fine spectra.

References

1. Ives, Herbert. Adventures in standing light waves. Rumford Medal Lecture 1951, Proceedings of the American Academy of Arts and Sciences, 1951, Vol. 81, No. 1, pp. 1-31.
2. Arya, Atam P. Introductory college physics. Macmillan, 1979, p. 750.
3. Barnes, Thomas G. Foundations of Electricity and Magnetism, D.C. Heath, 1965, pp. 264-267.

187

CHAPTER 15
Summary

The plausibility of a classical unification of physics has been developed for all of the physical forces of nature to be reduced to two fundamental forces, namely electric and magnetic forces. The classical approach to electrodynamics that introduced a new feedback concept and retained ordinary time and space (see Chapters 7 and 8) was perhaps the most significant development in making that unification possible. This new theory of electrodynamics opened the way for an alternative development of modern physics. One application of that theory assured the reasonableness of speeds exceeding the speed of light in certain cases, such as for example the rim speed of the electron or proton in their spinning state. That opened the way for a greater ratio of magnetic to electric field strength in the vicinity of the proton or electron than ever before considered possible. That in turn made possible a classical explanation of the so-called strong force in nuclear physics, a strong magnetic force caused by the high spin rate of the proton and electron.

Once the strong force of modern physics has been identified as a magnetic force, it opens the way for the development of classical models of the neutron, the hydrogen atom, the hydrogen molecule, and other atoms and molecules. This book has only given some simple plausible approaches to that development. The strong magnetic force provides the force re-

quired to bind two protons together in the nucleus, preventing
the electrostatic Coulomb repulsion from ejecting a proton
from the nucleus. It also provides the atomic force that prevents the electron from falling into the proton in the atom.
This magnetic force and the deformable property of the electron make possible structural configurations for the atom
that had never before been considered in classical physics. An
extension of these concepts suggested that the neutron consists of an electron and proton. The neutron is in unstable
equilibrium outside of the atom, a state from which it decays
into an electron and proton and emits its "excess" energy in
the form of electromagnetic radiation.

The rejection of the quantum postulate that an electron and
a proton can have only one value of spin, a fixed value, opens
the way for classical development of the concept of electromagnetic induction between the electron and proton. The
introduction of the postulate that the electron has no intrinsic
spin, only the spin produced by electromagnetic induction,
led to the logical conclusion that the electron is a *perfect diamagnetic body*. The additional postulate that the proton has
an intrinsic spin provides the assurance that there will always
be an induced spin in the electron in an atom. The diamagnetic property of the electron gives clues to the "allowed" positions of the electron in the atom and in the molecule. A diamagnetic body in a nonlinear magnetic field always has a repulsion force on it directed toward the weaker field. That is
an extremely important property of the electron's role in nuclear and atomic structures.

The *electric theory of inertial mass* was first developed for
the electron. The inertial property was shown to be the result
of an electromagnetically induced electric force acting backwards on the electron at the instant the electron is accelerated.
Its mass was then expressible in terms of the ratio of charge
squared to its radius. A logical extension of that derivation
led to the same equation for the mass of the proton. This
result indicated that the electron radius is larger than the pro-

ton radius by a factor of 1,836. After showing that the neutron consists of an electron and a proton, all mass is theoretically expressible in terms of electric charge.

Newton's third law was explained for the *dynamic* case from the point of view of the induced electric field acting backwards on the charges in the body when it is accelerated. The *static* case for Newton's third law was not developed in this book. However, the static reaction force must be a magnetic repulsion force. It is the result of the diamagnetic property of an electron. A reaction repulsion force is exerted when two bodies are forced together, close enough for the magnetic field from the protons and electrons to interact. In other words, the Newtonian reaction force in the static case is a diamagnetic repulsion force. Since there is no ohmic loss associated with the diamagnetic force of an electron, this reaction force is sustained as long as the static action force is applied.

The electric explanation of mass was deduced (see Chapter 3) without any reference to energy. *The so-called equivalence of mass and energy is a somewhat misleading concept.* Mass is one thing; energy is something else. An electric explanation was also given for the *moment of inertia*, the property associated with rotating masses. So the moment of inertia can also be expressed in terms of electric charge and the dimensions of the body. In theory it should be possible to reduce all of the equations of mechanics to electromagnetic equations. This classical unification of physics may be introduced into practical *engineering* applications.

In view of the work of Herbert Ives and of Charles Poor, the particle and mass concepts of light are rejected. Light does not have a dual nature. It is a wave phenomenon. It exerts pressure and propagates energy very much like an acoustic wave. It is not a ballistic phenomenon. Neither is it a billiard-ball-like particle that collides with elementary particles. It is believed that the experiments which have been conventionally interpreted that way can be reinterpreted in terms of wave interference phenomena.

In view of the electric theory of gravitation, a gravitational field does not attract light. This means that *there is no validity to the black hole theory*. Neither is there any observational evidence for black holes. Light has no mass because it has no electric charge. The gravitational force is an electric force acting on the electric charges within the mass.

The models for the neutron, hydrogen atom, and hydrogen molecule are tentative. The models may need to include the magnetic mass associated with the spin. That would alter the size, resonant frequencies, rim speeds and the associated magnetic forces. Much more work needs to be done to refine these models. Nevertheless, it is believed that this classical approach is a step in the right direction. It appears that it may soon be possible to show theoretically that resonant physical vibrations in the hydrogen model can radiate the known hydrogen spectral frequencies. It has already been shown that this model of the hydrogen atom can have *sufficient effective mass and high enough Q for natural resonant vibrational modes in the frequency range of the hydrogen spectrum*. The classical concept of electron to proton magnetic coupling employed in this model made that advance possible.

Glossary of Symbols

A	area
a	acceleration
B	magnetic induction field vector
C	capacitance
c	speed of light in vacuum
D	electric displacement field vector
E	electric field vector
F	force
f	frequency
G	universal gravitational constant
g	acceleration of gravity
H	magnetic field vector
h	Planck's constant
I	electric current
I	moment of inertia
J	electric current density (amp/m^2)
L	inductance
M	magnetic moment
m	mass
M	magnetization
N	number of protons
P	power
Q	sharpness of resonance
q	electric charge
R	resistance
r	radius or radial distance

S	spin angular momentum
s	distance
T	kinetic energy, magnetic energy, time constant
t	time
U	energy
V	electric potential, potential energy
v	velocity, speed
V	volume
W	energy
w	work or energy
α	angular acceleration
β	ratio of speed to speed of light
γ	relativistic increase factor
ϵ	permittivity
μ	permeability
ρ	charge density
τ	torque
ϕ	magnetic flux
ω	angular velocity in radian/sec
ν	frequency
Subscript e	electron
Subscript p	proton
Subscript n	neutron
Subscript m	magnetic

Appendix I

Physical Constants	Value
Speed of light (in vacuum)	2.9979×10^{8}
Universal gravitational constant	6.6732×10^{-11}
Planck constant	6.6262×10^{-34}
Electronic charge	1.6022×10^{-19}
Electron rest mass	9.1096×10^{-31}
Proton rest mass	1.6726×10^{-27}
Neutron rest mass	1.6749×10^{-27}
Bohr radius	5.2918×10^{-11}
Electron radius	1.8786×10^{-15}
Proton radius	1.0232×10^{-18}
Permeability μ (in vacuum)	$4\pi \times 10^{-7}$
Permittivity ϵ (in vacuum)	8.8542×10^{-12}
Electron magnetic moment	9.2849×10^{-24}
Proton magnetic moment	1.4106×10^{-26}
Nuclear magneton	5.0510×10^{-27}
Rydberg constant	1.0974×10^{7}
Hydrogen ionization potential	13.599 volt
Frequency of H_{α} line	$4.5680 \times 10^{14} \text{Hz}$

Appendix II
Inertial Mass of an Electric Current

By Dr. Russell Humphreys

3/13/83

MKS Units

Imagine a wire of radius a conducting a current I in the z-direction. It has a velocity $\vec{v}$ and an acceleration $\dot{\vec{v}}$, both in the x-direction. At time t=0 the center of the wire is at x=0.

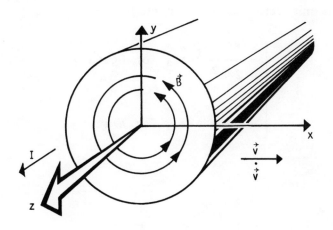

At a later time t, the center of the wire is at x=vt

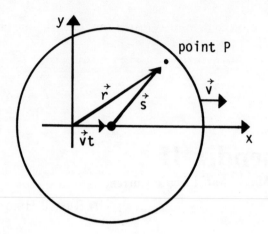

The vector $\vec{r}$ to the field point P stays fixed. The vector $\vec{s}$ from the wire center to P varies with time:

$$\vec{s} = \vec{r} - \vec{v}t . \qquad (1)$$

The magnetic field around the wire is

($\hat{z}$ is unit vector in z-direction, etc.)

$$\vec{B} = \begin{cases} -\dfrac{\mu_0 I}{2\pi a^2}\, \vec{s} \times \hat{z} , & \text{for } 0 \le s \le a , \qquad (2) \\[2em] -\dfrac{\mu_0 I}{2\pi s}\, \hat{s} \times \hat{z} , & \text{for } s > a . \qquad (3) \end{cases}$$

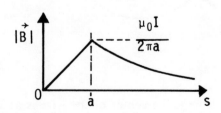

The rate of change of $\vec{B}$ *inside* the wire at time zero is:

$$\frac{\partial \vec{B}}{\partial t} = - \frac{\mu_0 I}{2\pi a^2} \frac{\partial \vec{s}}{\partial t} \times \hat{z} \quad , \text{ for } s \leq a \tag{4}$$

From Eq. (1) we can get:

$$\frac{\partial \vec{s}}{\partial t} = - \vec{v} \quad . \tag{5}$$

Using (5) in (4) gives:

$$\frac{\partial \vec{B}}{\partial t} = \frac{\mu_0 I}{2\pi a^2} \vec{v} \times \hat{z} = \frac{\mu_0 I}{2\pi a^2} v \hat{x} \times \hat{z} \tag{6}$$

Since $\hat{x} \times \hat{z} = -\hat{y}$, (6) becomes:

$$\frac{\partial \vec{B}}{\partial t} = - \frac{\mu_0 I}{2\pi a^2} v \hat{y} \quad , \quad s \leq a \tag{7}$$

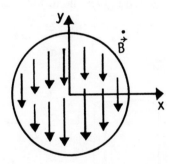

So $\frac{\partial \vec{B}}{\partial t}$ is the same everywhere within the wire, and in the negative y-direction. From Maxwell's first equation, the change of $\vec{B}$ induces an electric field $\vec{E}$:

$$\vec{\nabla} \times \vec{E} = - \frac{\partial \vec{B}}{\partial t} \quad . \tag{8}$$

Plugging (7) into (8) gives:

$$\vec{\nabla} \times \vec{E} = \frac{\mu_0 I}{2\pi a^2} v\hat{y} \ .$$

(9)

Expanding the curl in (9) gives:

$$\left(\frac{\partial E_x}{\partial z} - \frac{\partial E_z}{\partial x}\right)\hat{y} = \frac{\mu_0 I}{2\pi a^2} v\hat{y}$$

(10)

Since E_x is independent of z, $\dfrac{\partial E_x}{\partial z} = 0$, and (10) becomes:

$$\frac{\partial E_z}{\partial x} = - \frac{\mu_0 I}{2\pi a^2} v$$

(11)

Integrating (11) with respect to x and requiring that $\vec{E} = 0$ in the center of the wire gives the induced electric field $\vec{E}$ in the wire:

$$\vec{E} = - \frac{\mu_0 I}{2\pi a^2} vx\hat{z}$$

(12)

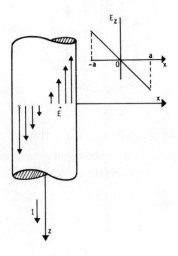

Let us say that the wire consists of many insulated strands in parallel. Then the induced electric field $\vec{E}$ will not produce any eddy currents in the wire. But since v is increasing with time, $\vec{E}$ is also increasing. Maxwell's second equation says that the changing $\vec{E}$ will produce an *induced* magnetic field $\vec{H}_i$ such that:

$$\vec{\nabla} \times \vec{H}_i = \varepsilon_0 \frac{\partial \vec{E}}{\partial t} .$$

(13)

Using (12) in (13) gives:

$$\vec{\nabla} \times \vec{H}_i = - \frac{\varepsilon_0 \mu_0 I}{2\pi a^2} \frac{\partial v}{\partial t} x\hat{z} .$$

(14)

Expanding the curl in Eq. (14) and using $\varepsilon_0 \mu_0 = \dfrac{1}{c^2}$ gives:

$$\left(\frac{\partial H_y}{\partial x} - \frac{\partial H_x}{\partial y}\right)\hat{z} = - \frac{\dot{I}vx}{2\pi a^2 c^2} \hat{z} .$$

(15)

One solution of (15) is

$$H_x = 0,$$

(16)

$$H_y = \frac{I\dot{v}}{4\pi c^2} \left(1 - \frac{x^2}{a^2}\right) .$$

(17)

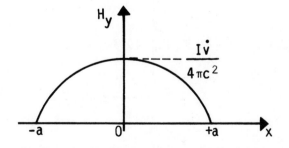

This meets the boundary condition of $\vec{H} = 0$ @ $x = \pm a$
(which comes from the fact that $\dfrac{d\vec{H}}{dx}$ must be continuous and
change sign at $x = \pm a$). So the induced magnetic field in the
wire is:

$$\vec{H}_i = \frac{I\dot{v}}{4\pi c^2}\left(1 - \frac{x^2}{a^2}\right)\hat{y} \ . \tag{18}$$

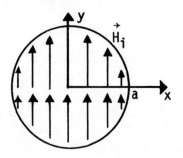

The field $\vec{H}_i$ produces a force on the current density $\vec{j}$ in
the wire. The Lorentz force per unit volume is:

$$\vec{f} = \vec{j} \times \left(\mu_0 \vec{H}_i\right) \ . \tag{19}$$

Since $\vec{j} = \dfrac{I}{\pi a^2}\hat{z}$, Eq. (19) becomes

$$\vec{f} = \frac{\mu_0 I}{\pi a^2}\hat{z} \times \vec{H}_i \ . \tag{20}$$

Plugging (18) into (20) and using the fact that $\hat{z} \times \hat{y} = -\hat{x}$

gives:

$$\vec{f} = - \frac{\mu_0 I^2}{4\pi^2 a^2 c^2} \left(1 - \frac{x^2}{a^2}\right) \dot{\vec{v}} \hat{x}$$ (21)

The force in Eq. (21) is opposite to the direction of acceleration, so it is a reaction force. The total $\vec{F}$ per unit length is

$$\vec{F} = - \frac{\mu_0 I^2}{4\pi a^2 c^2} \dot{\vec{v}} \hat{x} \int_{-a}^{+a} \int_{-\sqrt{a^2-y^2}}^{+\sqrt{a^2-y^2}} \left(1 - \frac{x^2}{a^2}\right) dx \, dy \ .$$ (22)

The integral works out to be $\frac{3}{4}\pi a^2$ (according to my calculations), which gives:

$$\vec{F} = - \frac{3\mu_0 I^2}{16\pi c^2} \dot{\vec{v}} \ .$$ (23)

So the *inertial mass* m per unit length is:

$$m = \frac{3\mu_0 I^2}{16\pi c^2} \ .$$ (24)

So it looks like an electric current does indeed have an inertial mass. We now need to see how this mass compares to the energy in the magnetic field. The magnetic energy U_m per unit length is:

$$U_m = \frac{1}{2} \int \vec{B} \cdot \vec{H} \, dA$$ (25)

Using cylindrical coordinates and equations (2) and (3) in (25) gives:

$$U_m \cong \frac{1}{2\mu_0} \int_0^a \frac{\mu_0^2 I^2}{4\pi^2 a^4} r^2 (2\pi r) dr$$

$$+ \frac{1}{2\mu_0} \int_a^b \frac{\mu_0^2 I^2}{4\pi^2 r^2} (2\pi r) dr \ .$$ (26)

Here b is an arbitrary radius of the order of the entire wire system, beyond which the $1/r$ field law no longer holds. Because of this limitation we cannot evaluate the field energy exactly, but we can get a rough estimate.

Integrating (26) then gives our estimate of magnetic field energy:

$$U_m \cong \frac{\mu_0 I^2}{16\pi} \left(1 + 4 \ln\frac{b}{a}\right) . \qquad (27)$$

Comparing the field energy (27) with the inertial mass (24) shows us that

$$m \cong \frac{U_m}{c^2} \text{ (order of magnitude only).} \qquad (28)$$

So in this case at least, we can conclude that:

(a) Electric currents have inertial mass.

(b) The mass is at least roughly equivalent to the energy of the magnetic field.

INDEX